Exercises on Geological Maps and DIP-Strike Problems

(For B.Sc. Students)

Exercises on Geological Maps and DIP-Strike Problems

GOKHALE

(For B.Sc. Students)

BY

DR. N.W. GOKHALE
RETIRED PROFESSOR OF GEOLOGY,
SHREYAS, 82 HUBLI ROAD,
DHARWAD 580007
KARNATAKA

CBS

CBS Publishers & Distributors Pvt. Ltd.

New Delhi • Bengaluru • Chennai • Kochi • Kolkata • Mumbai
Hyderabad • Nagpur • Patna • Pune • Vijayawada

ISBN: 81-239-0452-5

First Edition: 1996
Reprint: 2008, 2010, 2012, 2014, 2017, 2019

Published by **Satish Kumar Jain** and produced by **Varun Jain** for
CBS Publishers & Distributors Pvt. Ltd.,
4819/XI Prahlad Street, 24 Ansari Road, Daryaganj, New Delhi - 110002
delhi@cbspd.com, cbspubs@airtelmail.in • www.cbspd.com
Ph.: 23289259, 23266861, 23266867 • Fax: 011-23243014

Corporate Office: 204 FIE, Industrial Area, Patparganj, Delhi - 110 092
Ph: 49344934 • Fax: 011-49344935
E-mail: publishing@cbspd.com • publicity@cbspd.com

Branches:
- ***Bengaluru:*** 2975, 17th Cross, K.R. Road, Bansankari 2nd Stage, Bengaluru - 70 • Ph: +91-80-26771678/79 • Fax: +91-80-26771680 E-mail: cbsbng@gmail.com, bangalore@cbspd.com
- ***Chennai:*** No. 7, Subbaraya Street, Shenoy Nagar, Chennai - 600030 Ph: +91-44-26681266, 26680620 • Fax: +91-44-42032115 E-mail: chennai@cbspd.com
- ***Kochi:*** Ashana House, 39/1904, A.M. Thomas Road, Valanjambalam, Ernakulum, Kochi • Ph: +91-484-4059061-65 Fax: +91-484-4059065 • E-mail: cochin@cbspd.com
- ***Kolkata:*** 6-B, Ground Floor, Rameshwar Shaw Road, Kolkata - 700014 Ph: +91-33-22891126/7/8 • E-mail: kolkata@cbspd.com
- ***Mumbai:*** 83-C, Dr. E. Moses Road, Worli, Mumbai - 400018 Ph: +91-9833017933, 022-24902340/41 • E-mail: mumbai@cbspd.com

Representatives:
- Hyderabad: 0-9885175004
- Nagpur: 0-9021734563
- Patna: 0-9334159340
- Pune: 0-9623451994
- Jharkhand: 0-9811541605
- Uttarakhand: 0-9716462459

Printed at: J.S. Offset Printers, Delhi (India)

FOREWORD

Geological maps and drawing sections through appropriate lines to reveal the underground structure is one of the important aspects of structural geology. The geological structure of any region is the foundation for further geological studies. As a first step, the beginner is taught the method of drawing geological section along a suitable line to reveal the underground structure. Not many books have been published for providing exercises for undergraduate students. The manual "Exercises in Geological Maps and Dip and Strike problems" brought out by Prof. Gokhale, formerly Professor and Head of Geology Department, Karnatak University, Dharwad, is a welcome addition and will go a long way to fulfill the need.

The Manual will be found useful for undergraduate students of geology. The exercises span wide range from very simple to those involving folding, faulting and intrusions. The maps can be removed from the manual for section-drawing during the practicals.

The manual will also be very useful to civil engineering students.

P.S. AGARKAR
Formerly Professor and Head
Geology Department
Government Science College Jabalpur.

PREFACE

This is a Journal *compiling* 5 topographical maps, 15 geological maps (covering horizontally reposing formations, dipping beds, folded beds, faults and dyke intrusions), 10 maps for tracing of outcrops of beds (covering horizontally reposing, and dipping formations), and 35 Dip-Strike Problems (which cover 7 varieties). Besides this, there are additional exercises on topographical and geological maps, which is an unique feature of this journal. To obtain the accurate strike interval (this pertains to the maps involving tracing of outcrops) corresponding to the given amount of true dip, multiply it by the value of contour interval stated for the concerned map. Thus in the case of Map 36, multiply 6 (this being the given amount of true dip) by 25 (this being the given contour interval) which gives the strike interval of 150 meters. The strike lines are to be drawn 1.5 cms. apart from one another (because the given horizontal scale is 1 cm. =100 Meters).

In order to solve the exercises set in this journal, "A Manual of Geological Maps, and Problems in Practical Structural Geology" written by me, will be found indispensible. These two books will be of great help in solving maps and problems set out in this journal. The journal is designed specially for the students of B.Sc. Geology. It can be used by students of any university.

DR. N.W. GOKHALE
Retired University Professor
of Geology, Karnatak University,
Dharwad—3.

Dharwad—7

CONTENTS

Foreword .... *v*

Preface .... *vii*

SECTION - I : TOPOGRAPHY 1

Map 1. Isolated hillock with uniform slope on all sides. 3

Map 2. Isolated depression/basin with uniform slope on all sides. 5

Map 3. Elongated hillock with non-uniform slope on sides. 7

Map 4. Elongated depression/basin with non-uniform slope on sides. 9

Map 5. Several isolated hills and radial valleys. 11

Additional exercises on topographical maps 13

SECTION - II : HORIZONTAL BEDS 15

Map 6. Horizontal beds. 17

Map 7. Horizontal beds traversed by a vertical dyke. 19

Map 8. Horizontal beds affected by a vertical fault. 21

Map 9. Horizontal beds affected by two parallel vertical faults producing step faults. 23

Map 10. Horizontal beds affected by a concealed fault, a vertical dyke occupying the fault. 25

Additional exercises on horizontal beds 27

SECTION - III : DIPPING BEDS 29

Map 11. Dipping series of beds exposed in a hilly terrane. 31

Map 12. Dipping series of beds exposed in two depressions separated by ahillock. 33

Map 13. Dipping series of beds exposed on a sloping ground. Three point method of drawing initial strike line. 35

Map 14. Dipping series of beds traversed by a vertical dyke. Note the elongation of hill due to the vertical dyke. 37

Map 15. Dipping series of beds traversed by an inclined dyke. Note that the outcrop lines of the dyke are slightly curved as compared to those of the beds. This indicates that the dyke is more steeply inclined than the beds. 39

Map 16. Dipping series of beds traversed by two vertical intersecting dykes differing in age. Study the intersection area of the dykes, carefully. Basaltic dyke is younger than the dolerite dyke, because its outcrop lines do not stop against those of dolerite. 41

Map 17. Dipping series of beds traversed by two vertical intersecting dykes of same age. Study the area around the intersection of the two dykes critically, and compare it with that observed in Map 16. 43

Map 18. Dipping series of beds traversed by two intersecting dykes differing in age and attitude. The basaltic dyke is inclined and younger than the dolerite dyke. 45

Map 19. Dipping series of beds affected by a vertical dip fault. Note the shift in the outcrops along the fault. 47

Map 20. Dipping series of beds affected by a vertical strike fault. 49

Map 21. Dipping series of beds affected by a vertical oblique fault. 51

Map 22. Dipping series of beds affected by two vertical, parallel faults producing step faults. 53

Map 23. Dipping older series of beds laid over by younger horizontal beds, producing unconformity. 55

Map 24. Two dipping series of beds separated by an angular unconformity. 57

Map 25. Folded series of beds. Note the symmetrical nature of the folded structure. 59

Map 26. Folded series of beds. Note the asymmetrical nature of the folded structure. 61

Map 27. Folded series of beds. Count the number of anticlines and synclines. 63

Map 28. Folded series of beds with unconformity. Note the changes in the shapes of the anticline and syncline. 65

Map 29. Folded series of beds affected by a diagonal/oblique fault, and traversed by younger vertical dyke. Note the shift of the outcrop lines along the fault line. 67

Map 30. A horizontal series of beds traversed by three vertical dykes differing in age. Study the places of intersection of the dykes very carefully, and establish the relative ages. 69

Additional Exercises on Dipping Beds **71**

SECTION-IV : TRACING OF OUTCROPS, i.e., CONSTRUCTION OF GEOLOGICAL MAPS **73**
HORIZONTAL BEDS : INCLINED BEDS

HORIZONTAL BEDS

Map 31. Initial outcrop point given on contour by way of a cross. 75

Map 32. Initial outcrop point stated in the text. 77

INCLINED BEDS **79**

Map 33. Data given in the form of a continuous line. Note that two consecutive basic strike lines are readily obtainable. 79

Map 34. Data given in the form of a continuous line on two consecutive contour lines. Note that Two basic strike lines are immediately available from which the amount of dip and direction can be calculated. 81

Map 35. Data given in the form of a continuous line but not on two consecutive contour lines. Note that a strike line is required to be introduced in between the two strike lines initially drawn. Only then the correct amount of dip can be obtained. 83

Map 36. Data given in the form of true dip amount and direction, and true strike in the text, for a point marked by +/- on one contour. This gives only one strike line, others are to be drawn considering the given amount of dip. 85

Map 37. Data given in the text for two points marked by +/- on one contour, in the form of dip amount and general direction of dip. This gives only one basic strike line. Others are to be drawn considering the given amount of dip. 87

Map 38. Four places marked by +/- on same contour are given. These points give the first basic strike line. Others are to be drawn with the help of amount of dip and its direction, as given in the text. 89

SECTION - V : DIP AND STRIKE PROBLEMS **91**

Categories

A. Given two apparent dips, determine true dip amount, and direction of true dip and true strike. 93

B. Given true dip, determine dip amount in the given directions of apparent dip. 93

C. Given one apparent dip and true direction of strike, find true dip amount and direction. 94

D. Given one apparent dip, direction of true dip, find amount of true dip and strike. 94

E. Given one apparent dip and amount of true dip, find the direction of true dip. 94

F. Given true dip, to find the directions in which the given amount of apparent dip will be encountered. 95

G. Given one apparent dip and true dip and direction of true strike. 95

SECTION--I
TOPOGRAPHY

Map **1.** Isolated hillock with uniform slope on all sides.

Map **2.** Isolated depression/basin with uniform slope on all sides.

Map **3.** Elongated hillock with non-uniform slope on sides.

Map **4.** Elongated depression/basin with non-uniform slope on sides.

Map **5.** Several isolated hills and radial valleys.

Additional exercises on topographical maps

Map 1.

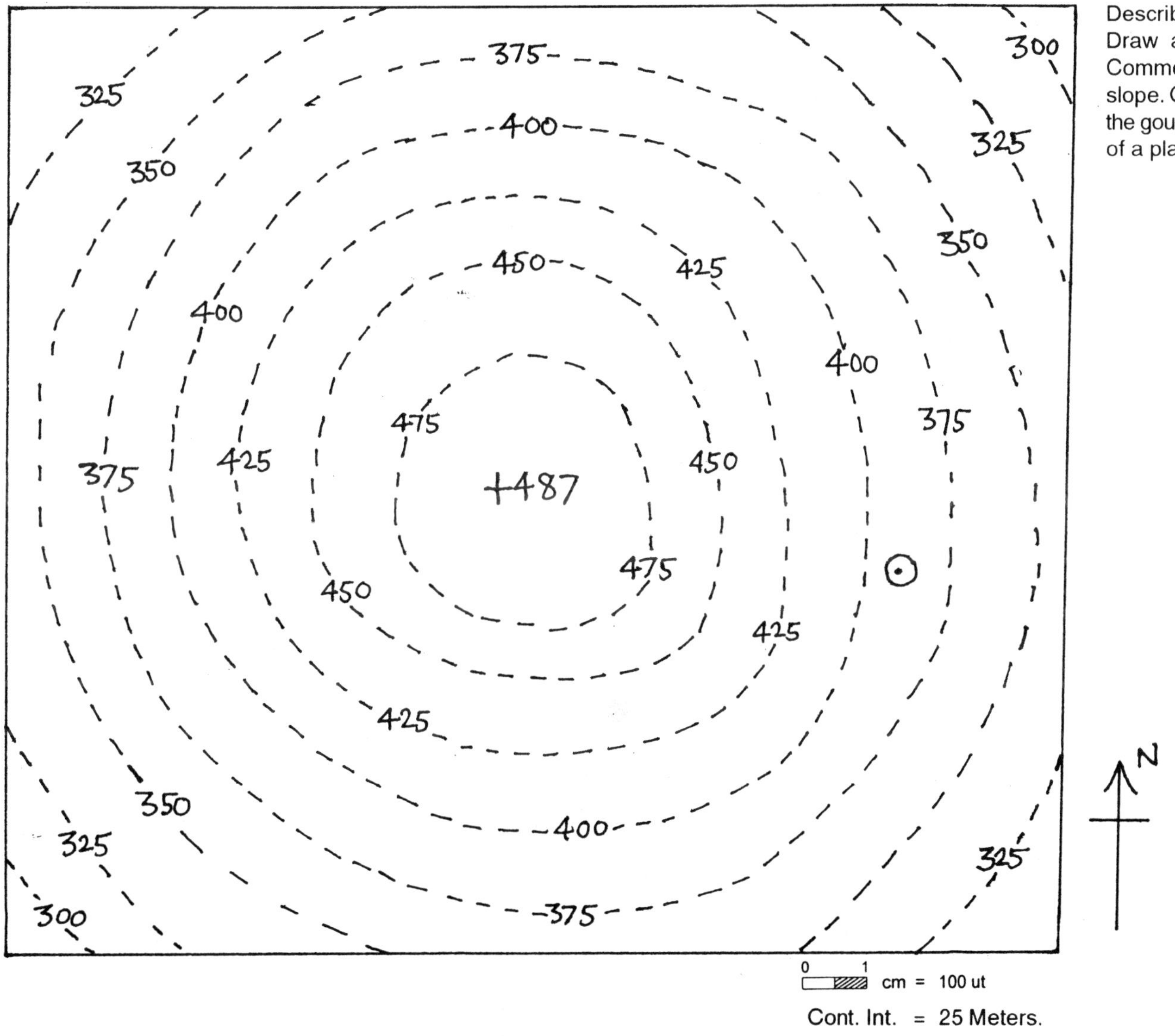

Describe topography of the area. Draw a suitable cross section. Comment on the nature of the slope. Calculate average slope of the gound surface. Estimate height of a place marked.

Map 2.

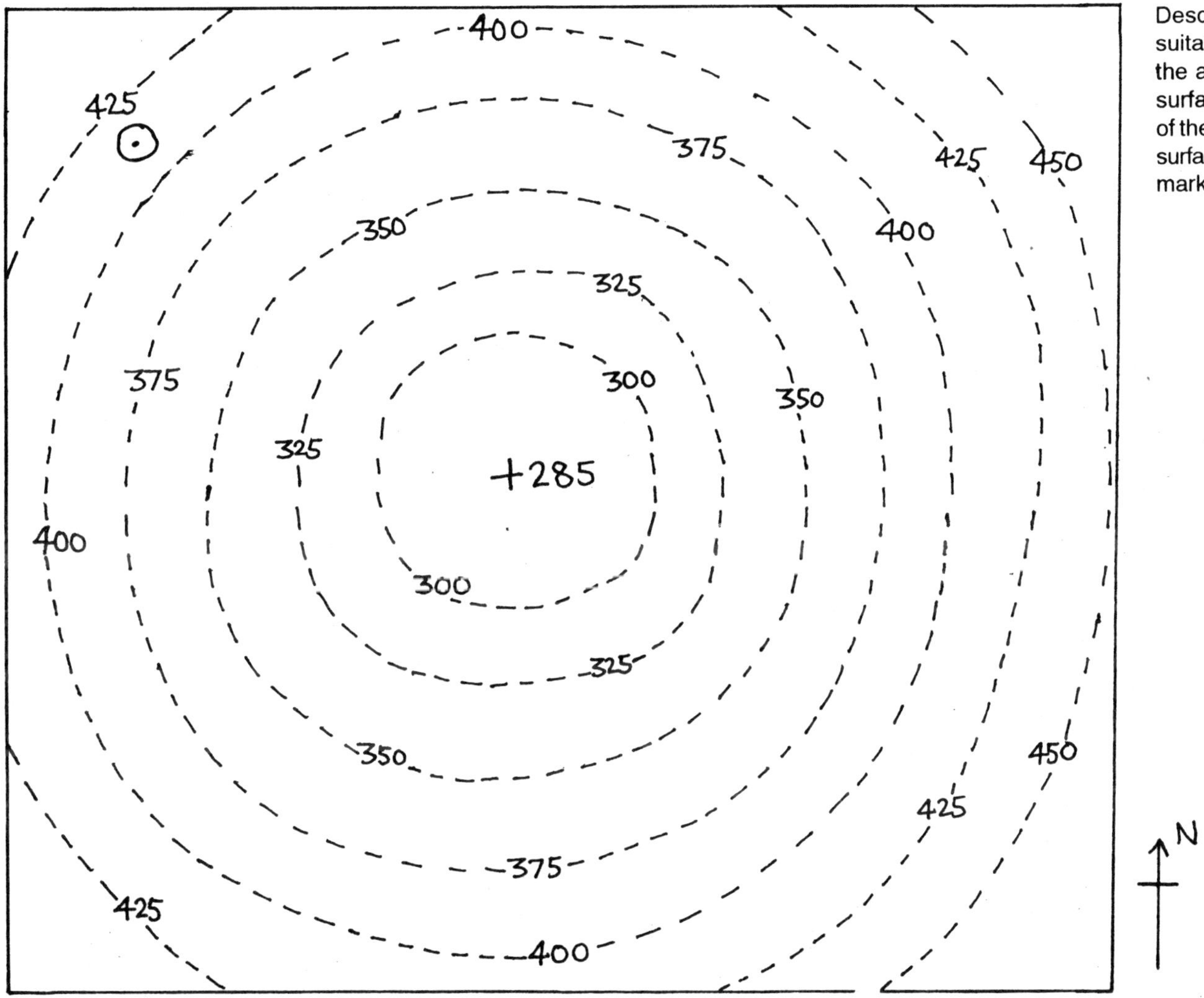

Describe the topography. Draw a suitable cross section. Calculate the average slope of the ground surface. Comment on the nature of the general slope of the ground surface. Estimate height of a place marked.

Map 3.

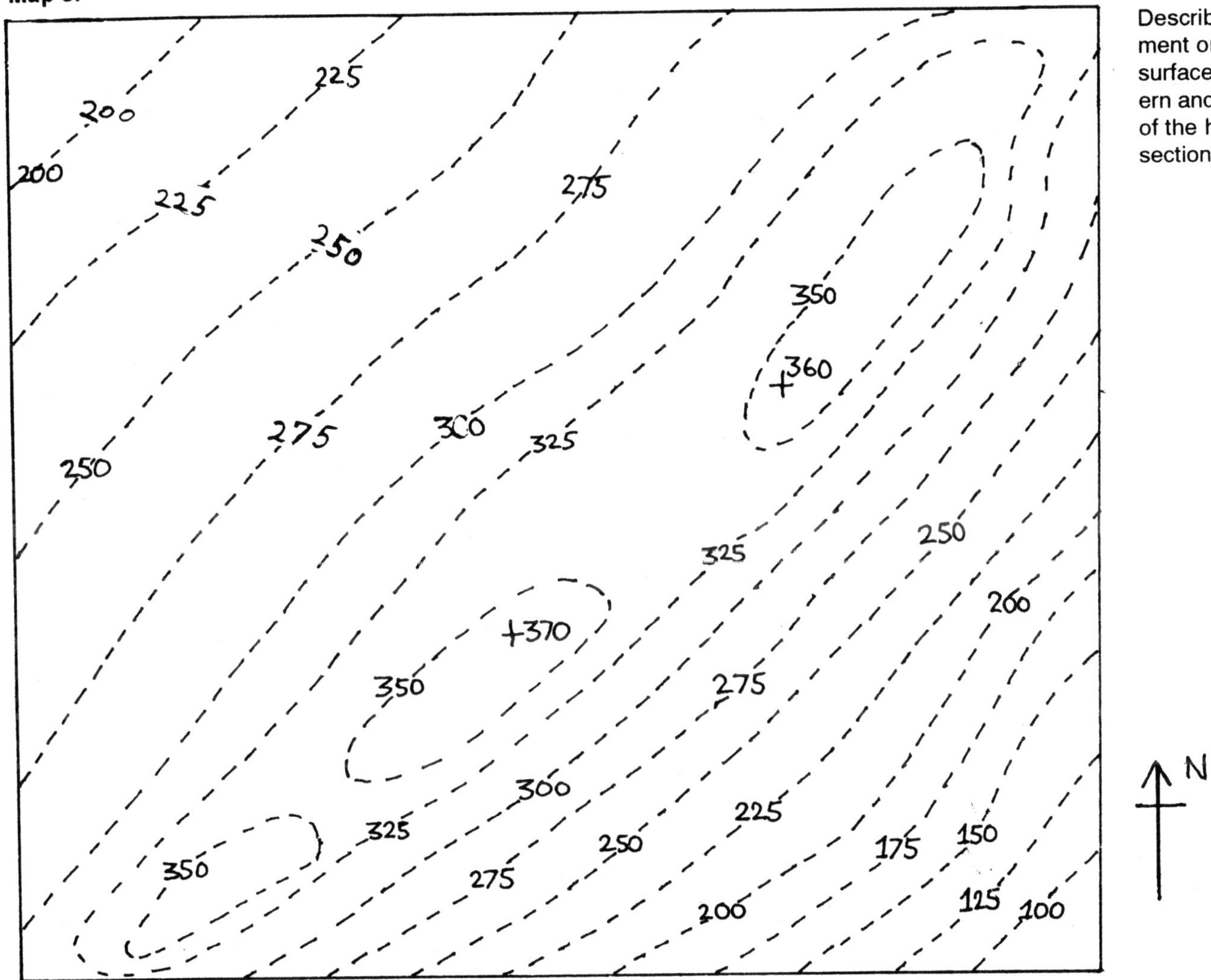

0 1 cm = 100 ut

Cont. Int = 25 Meters

Describe the topography. Comment on the slope of the ground surface. Calculate the north western and the south eastern slopes of the hill. Draw a suitable cross section.

Map 4.

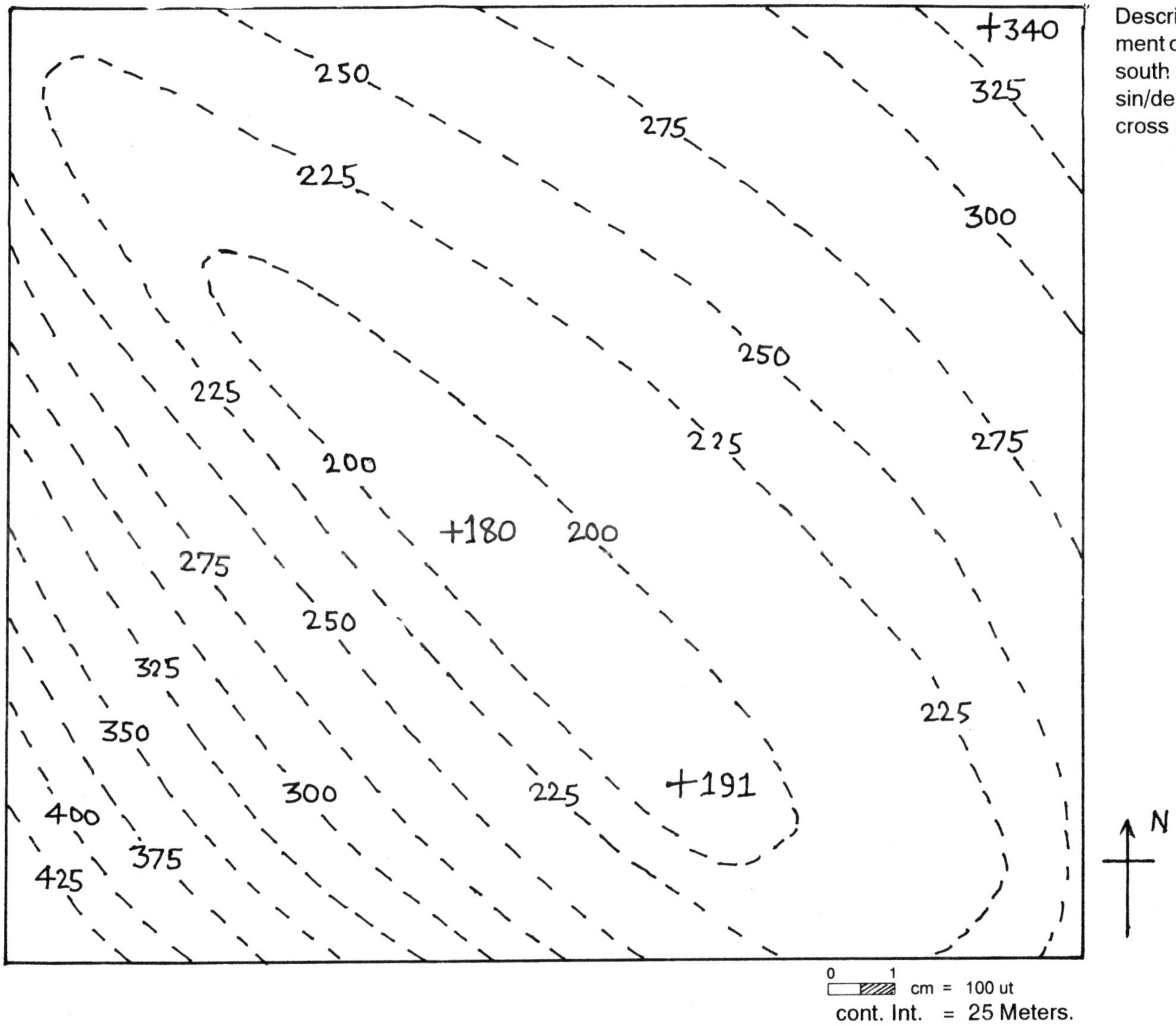

Describe the topography. Comment on the north eastern and the south western slopes of the basin/depression. Draw a suitabe cross section.

Map 5.

Describe topography. Comment on the valleys. Draw a suitable cross section.

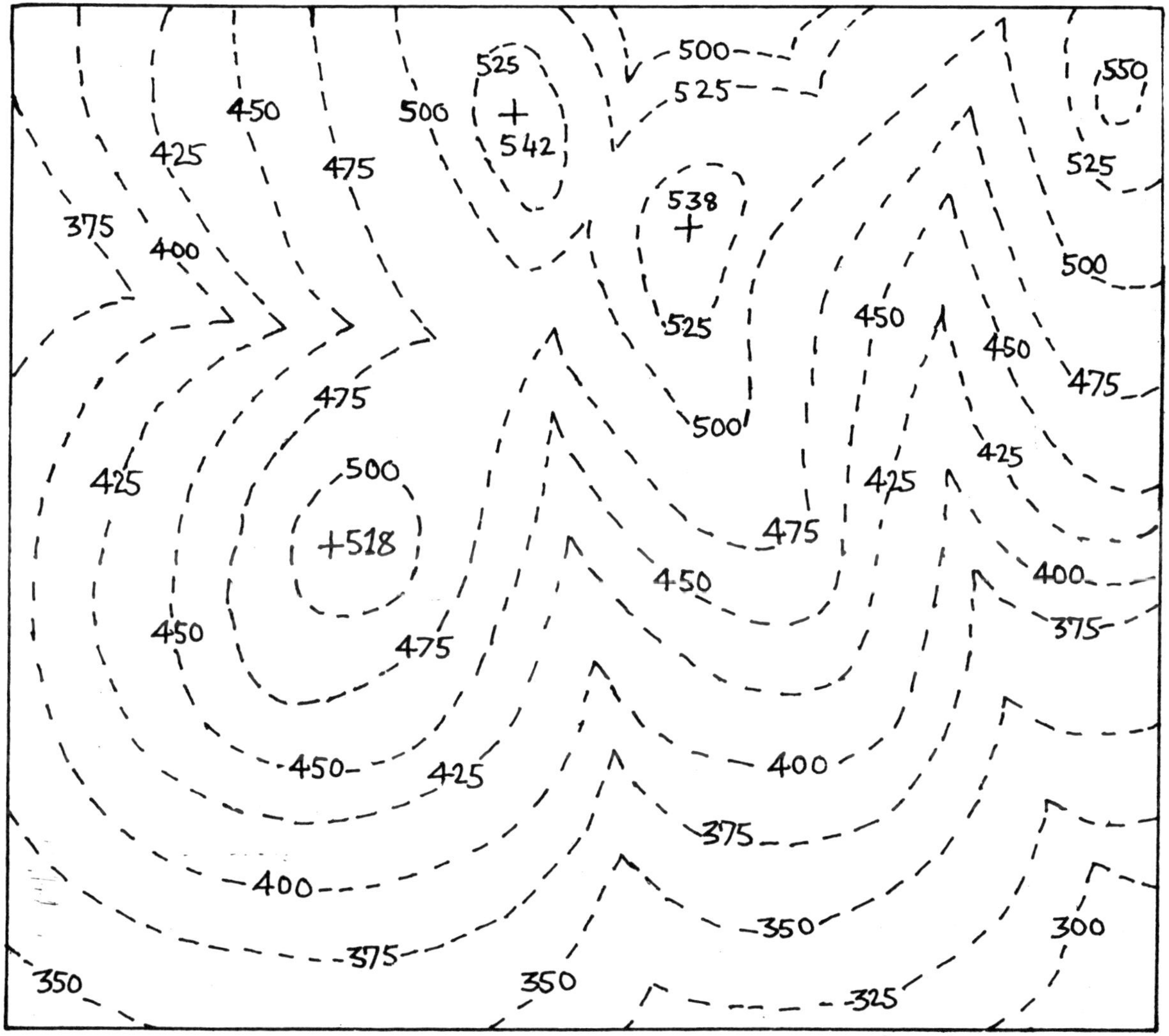

0 1 cm = 100 ut

Cont. Int = 25 Meters.

Additional Exercises on Topographical Maps

A. Draw a contour map of a hill such that its northern slope is steeper than the southern one.

B. Draw a contour map wherein there are isolated hillocks on the northwestern and the southeastern parts, and a valley is running in between the two hillocks.

C. Draw a contour map of a NW-SE trending valley with an isolated hill only on the northern side.

SECTION--II
HORIZONTAL BEDS

Map 6. Horizontal beds.

Map 7. Horizontal beds traversed by a vertical dyke.

Map 8. Horizontal beds affected by a vertical fault.

Map 9. Horizontal beds affected by two parallel vertical faults producing step faults.

Map 10. Horizontal beds affected by a concealed fault, a vertical dyke occupying the fault.

Additional exercises on horizontal beds

Map 6.

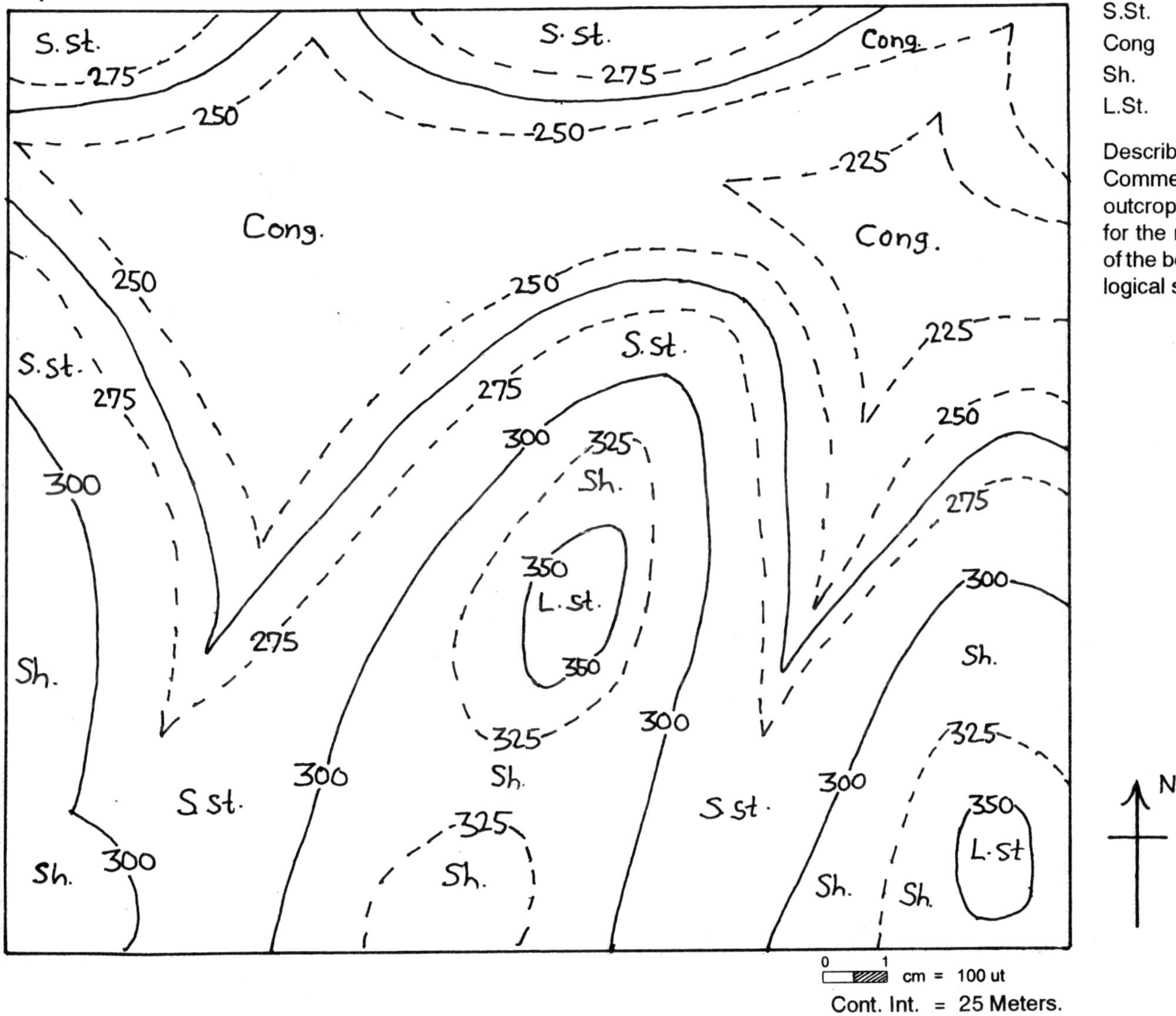

S.St. = Sandstone.
Cong = Conglomerate.
Sh. = Shale.
L.St. = Limestone.

Describe geology of the area. Comment on the shapes of the outcrops of the beds. Give reason for the repetition of the outcrops of the beds. Draw a suitable geological section.

Map 7.

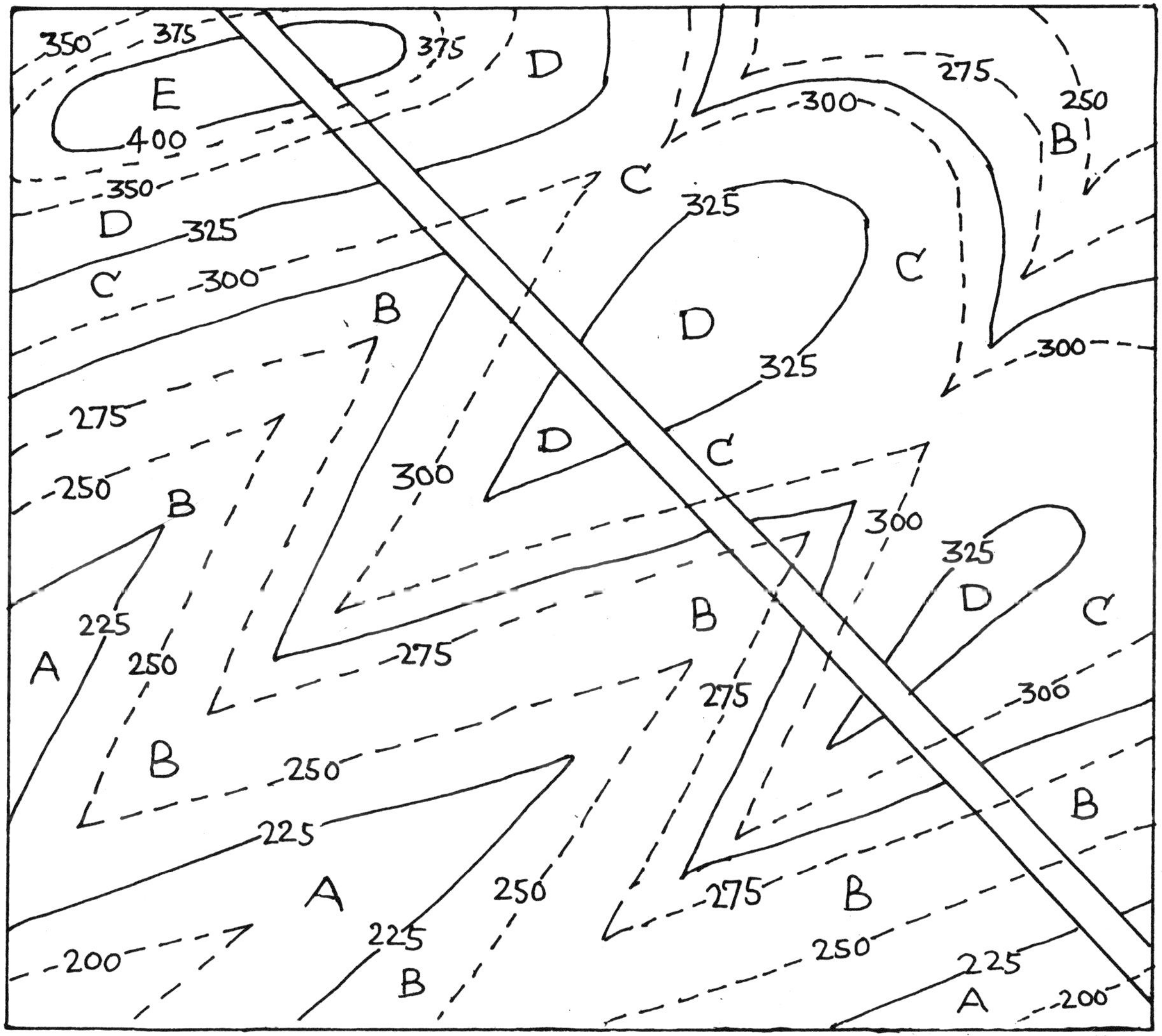

Describe geology of the area. Draw a suitable geological section. Comment on the shapes of the outcrops and the morphology. Comment on the repetition of the bed B in the northeastern corner of the area. Study the relation of dyke with the other beds.

0 1 cm = 100 ut

Cont. Int. = 25 Meters.

Map 8.

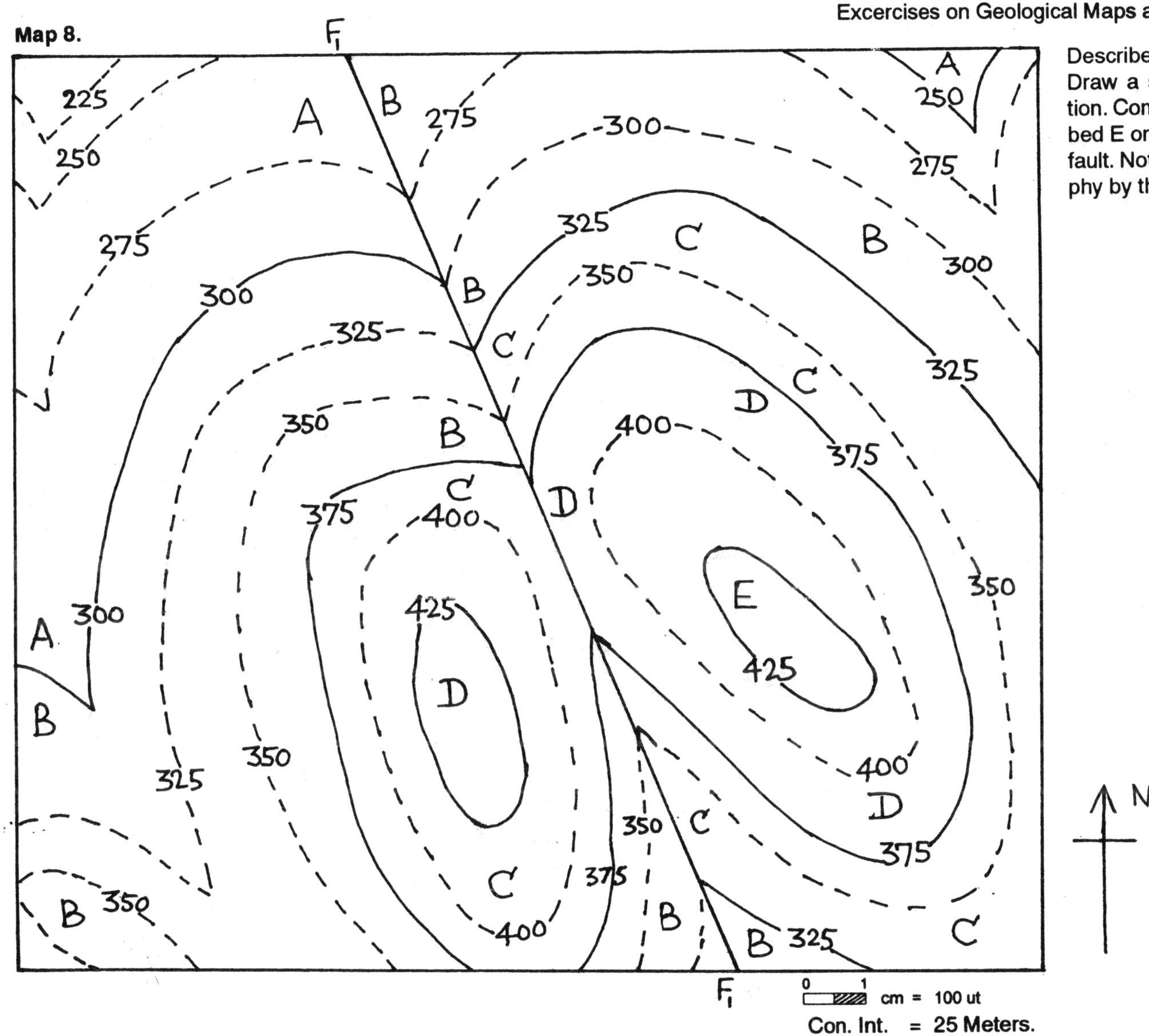

Describe the geology of the area. Draw a suitable geological section. Comment on the absence of bed E on the western side of the fault. Note the control of topogrophy by the fault.

Map 9.

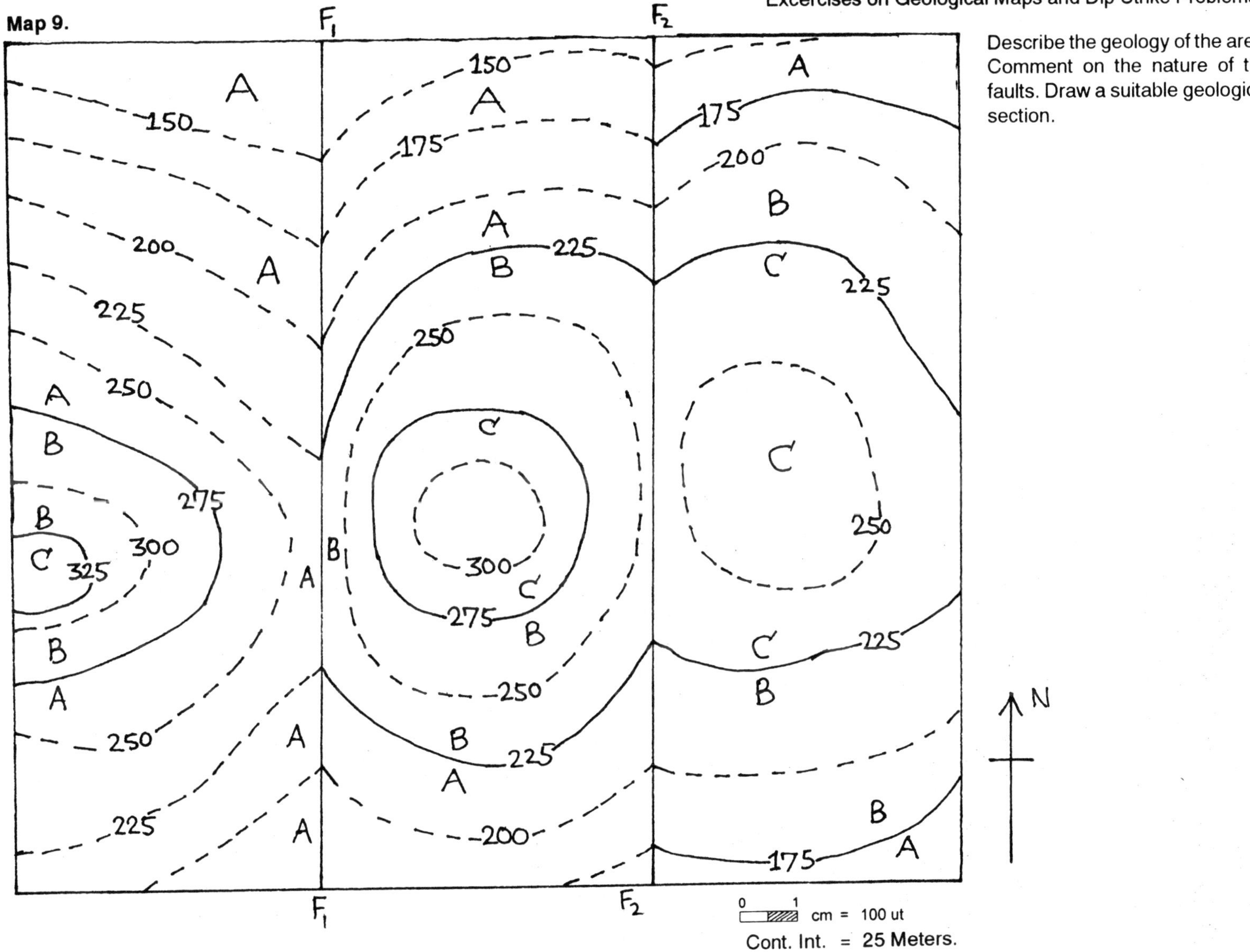

Describe the geology of the area. Comment on the nature of the faults. Draw a suitable geological section.

Map 10.

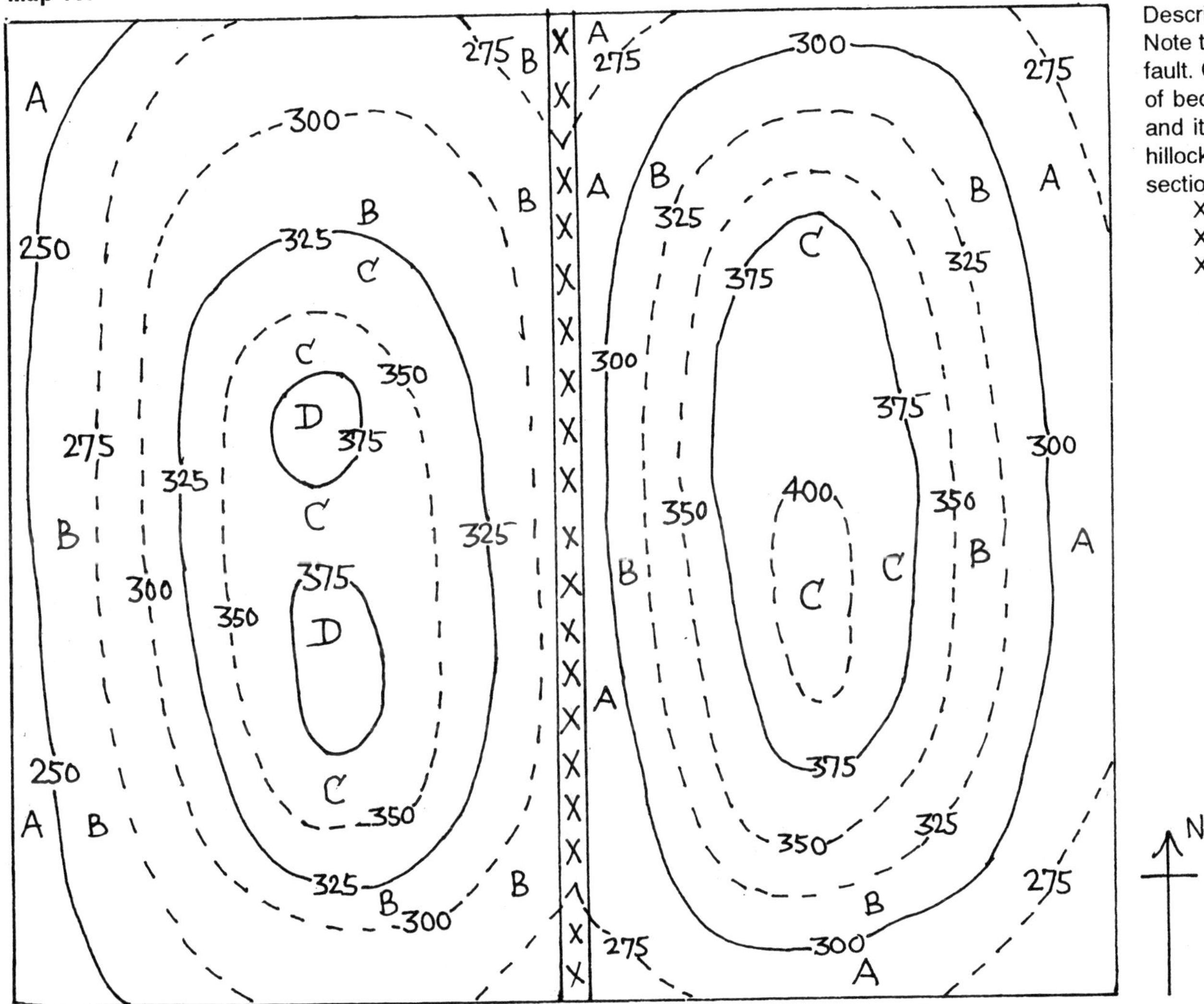

Describe the geology of the area. Note that the dyke conceals the fault. Comment on the repetition of bed D on the western hillock and its absence on the eastern hillock. Draw a suitable geological section.

X
X Dyke.
X

0 1 cm = 100 ut

Cont. Int. = 25 Meters.

Additional Exercises on Horizontal Beds

A. How is the horizontality of a bed determined?

B. What controls shapes of outcrops of beds?

C. How is the thickness of a bed determined?

D. What is the shape of outcrop of a vertical bed/dyke?

E. How is the thickness of a vertical dyke determined?

F. Note that the outcrops of the horizontal beds are repeated due to

   (i) Favourable topography and or

   (ii) Faulting.

G. Note that the outlier is formed only on the top of a hill, while an inlier is developed only in the valley portions or low lying areas.

SECTION--III
DIPPING BEDS

Map 11. Dipping series of beds exposed in a hilly terrane.

Map 12. Dipping series of beds exposed in two depressions separated by ahillock.

Map 13. Dipping series of beds exposed on a sloping ground. Three point method of drawing initial strike line.

Map 14. Dipping series of beds traversed by a vertical dyke. Note the elongation of hill due to the vertical dyke.

Map 15. Dipping series of beds traversed by an inclined dyke. Note that the outcrop lines of the dyke are slightly curved as compared to those of the beds. This indicates that the dyke is more steeply inclined than the beds.

Map 16. Dipping series of beds traversed by two vertical intersecting dykes differing in age. Study the intersection area of the dykes, carefully. Basaltic dyke is younger than the dolerite dyke, because its outcrop lines do not stop against those of dolerite.

Map 17. Dipping series of beds traversed by two vertical intersecting dykes of same age. Study the area around the intersection of the two dykes critically, and compare it with that observed in Map 16.

Map 18. Dipping series of beds traversed by two intersecting dykes differing in age and attitude. The basaltic dyke is inclined and younger than the dolerite dyke.

Map 19. Dipping series of beds affected by a vertical dip fault. Note the shift in the outcrops along the fault.

Map 20. Dipping series of beds affected by a vertical strike fault.

Map 21. Dipping series of beds affected by a vertical oblique fault.

Map 22. Dipping series of beds affected by two vertical, parallel faults producing step faults.

Map 23. Dipping older series of beds laid over by younger horizontal beds, producing unconformity.

Map 24. Two dipping series of beds separated by an angular unconformity.

Map 25. Folded series of beds. Note the symmetrical nature of the folded structure.

Map 26. Folded series of beds. Note the asymmetrical nature of the folded structure.

Map 27. Folded series of beds. Count the number of anticlines and synclines.

Map 28. Folded series of beds with unconformity. Note the changes in the shapes of the anticline and syncline.

Map 29. Folded series of beds affected by a diagonal/oblique fault, and traversed by younger vertical dyke. Note the shift of the outcrop lines along the fault line.

Map 30. A horizontal series of beds traversed by three vertical dykes differing in age. Study the places of intersection of the dykes very carefully, and establish the relative ages.

Map 11.

Describe geology of the area. Draw a suitable geological section.

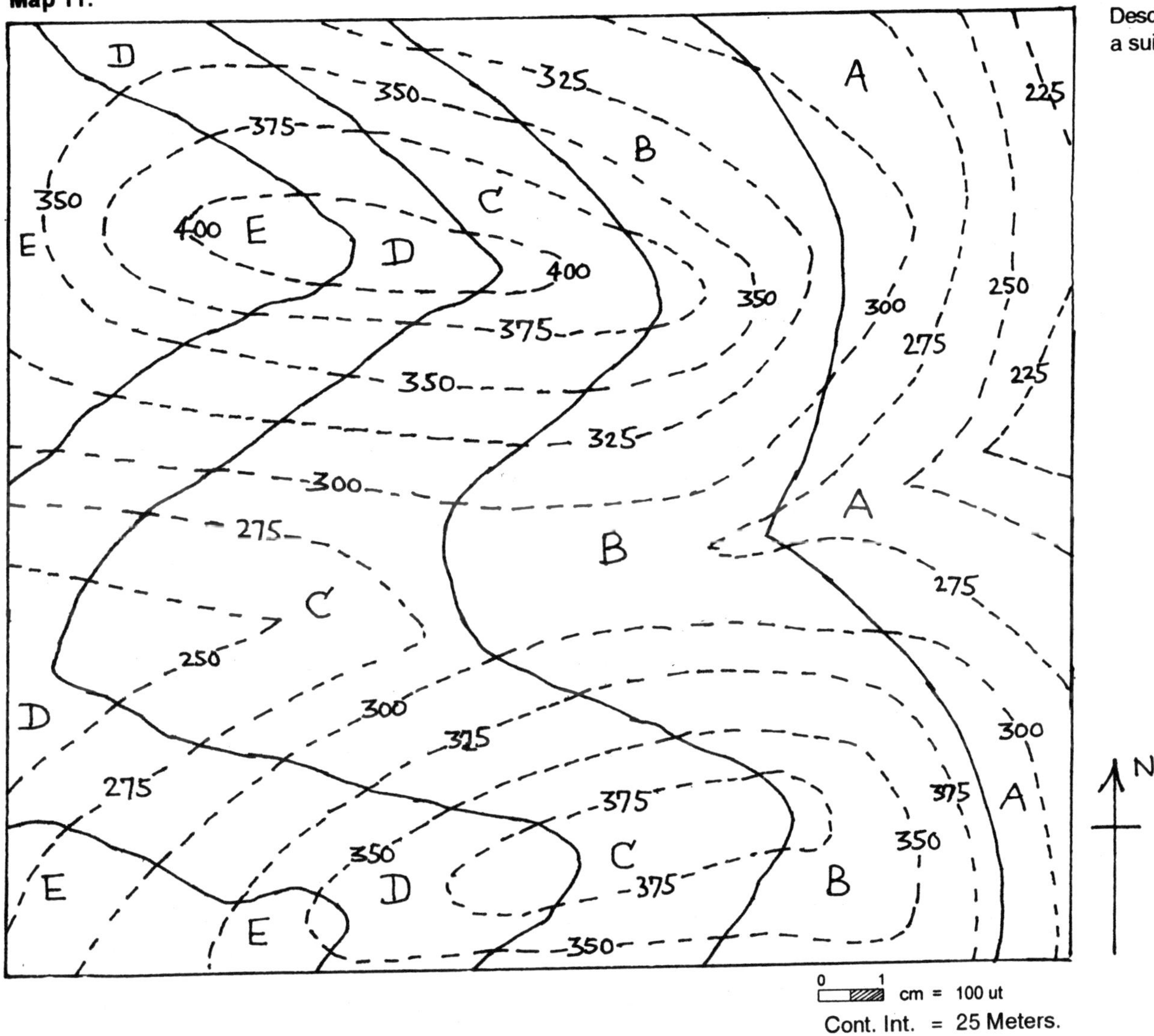

Map 12.

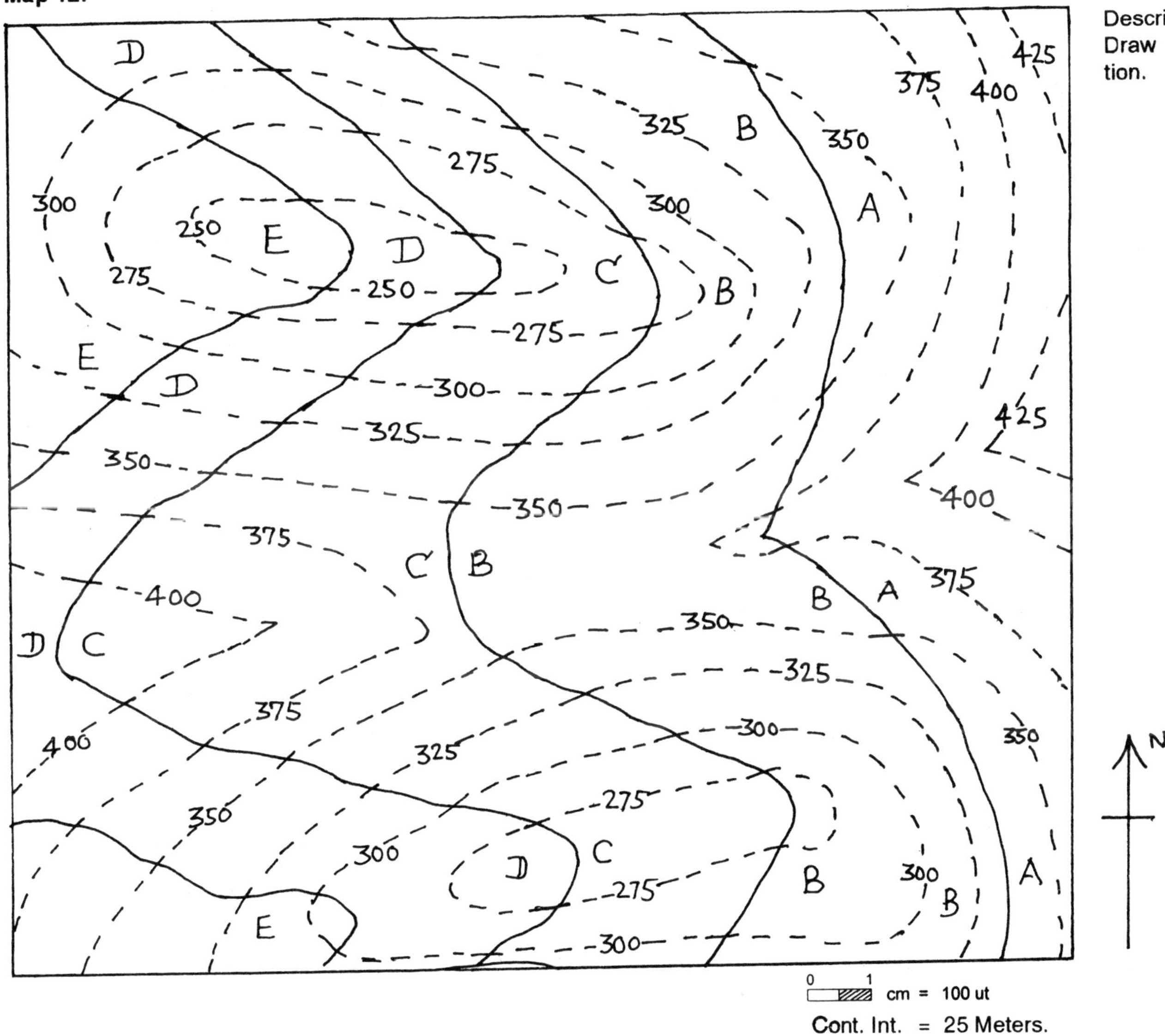

Describe the geology of the area. Draw a suitable geological section.

Map 13.

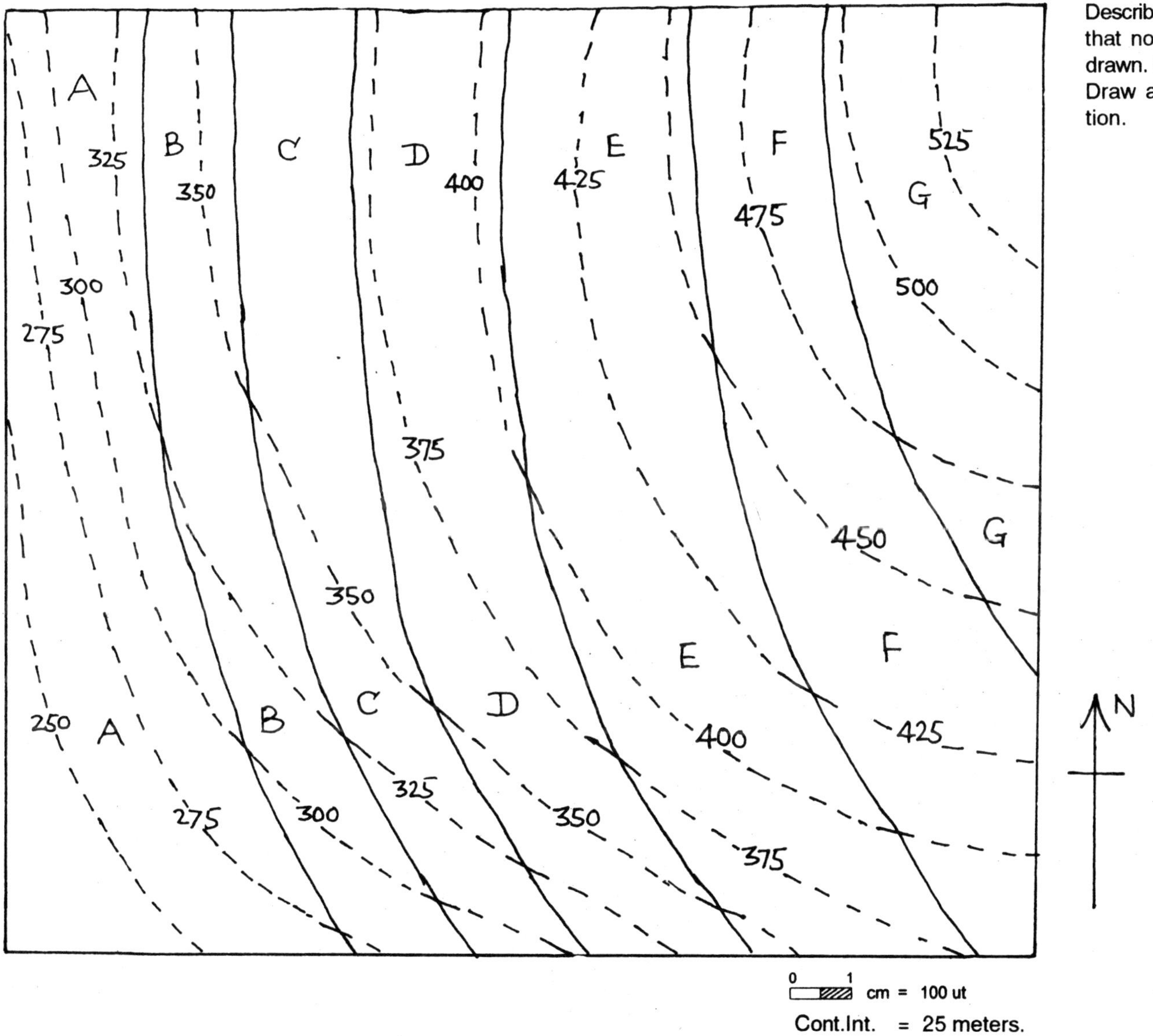

Describe geology of the area. Note that no direct strike line can be drawn. Follow three point method. Draw a suitable geological section.

Map 14.

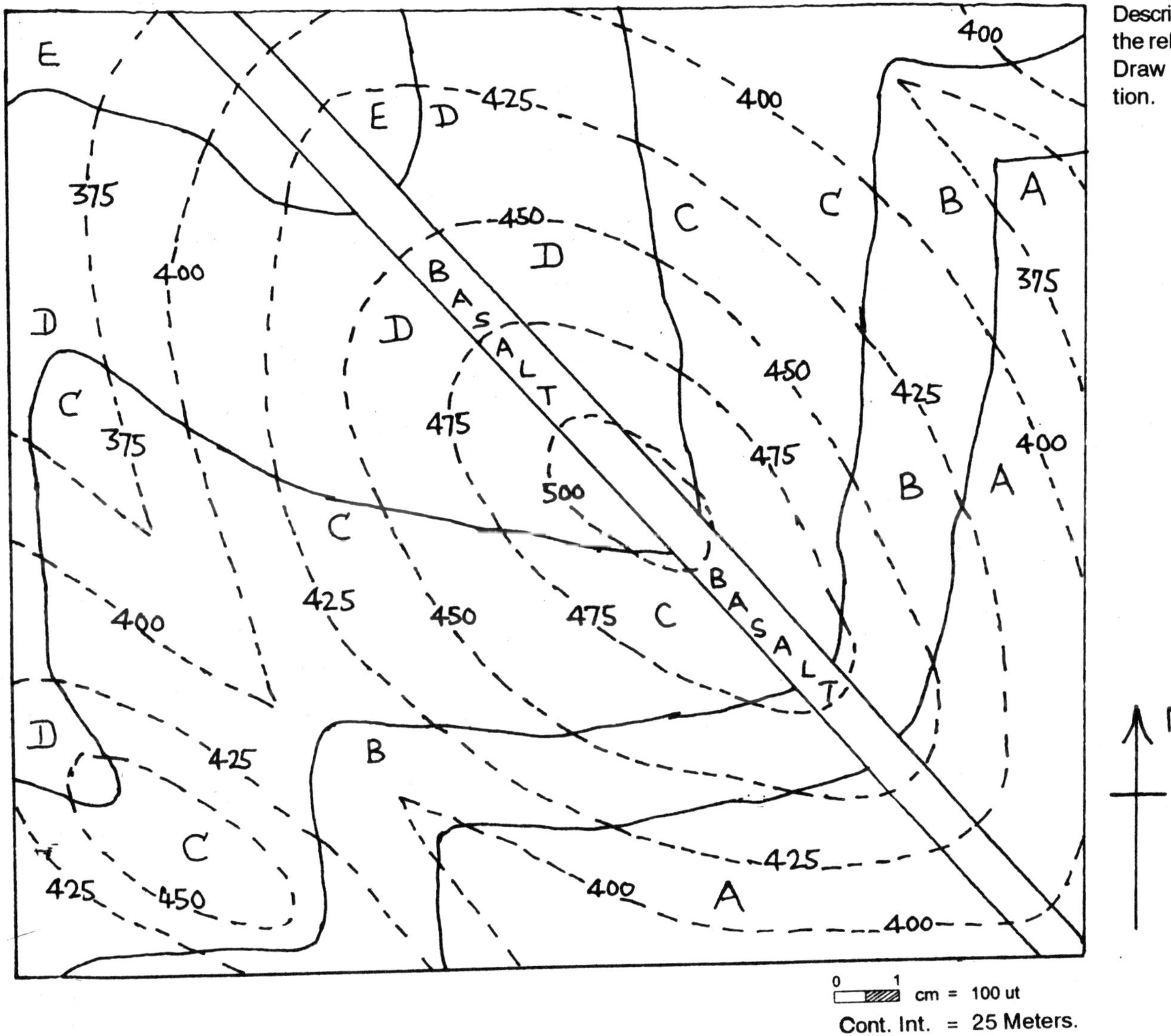

Describe geology of the area. Study the relation of dyke with the beds. Draw a suitable geological section.

Map 15.

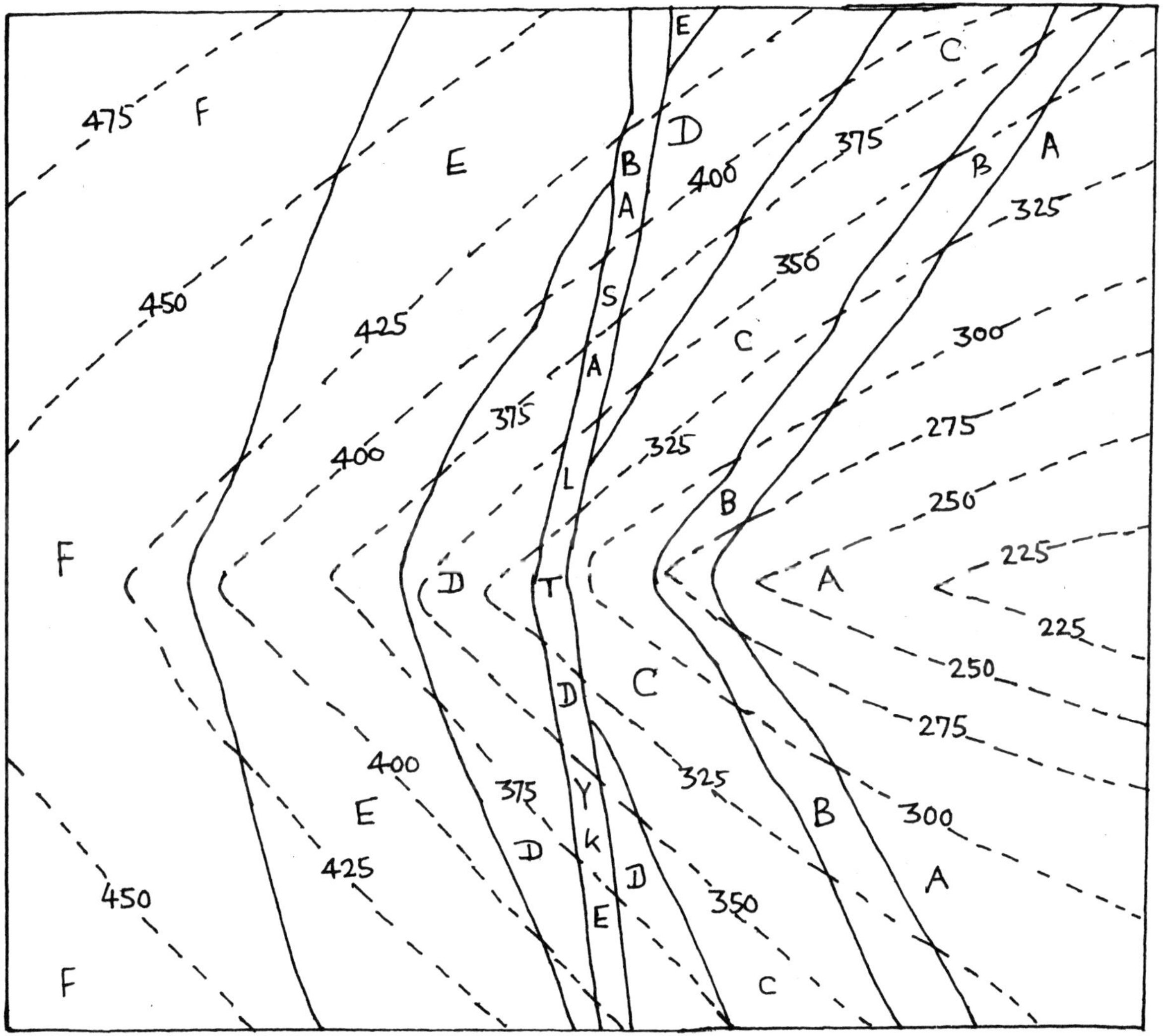

Describe the geology of the area. Note the marked difference between the curvatures of the outcrops of the dyke and the other beds. Draw separate strike lines for the inclined dyke. Draw a suitable geological section.

0 1 cm = 100 ut

Cont. Int. = 25 Meters.

Map 16.

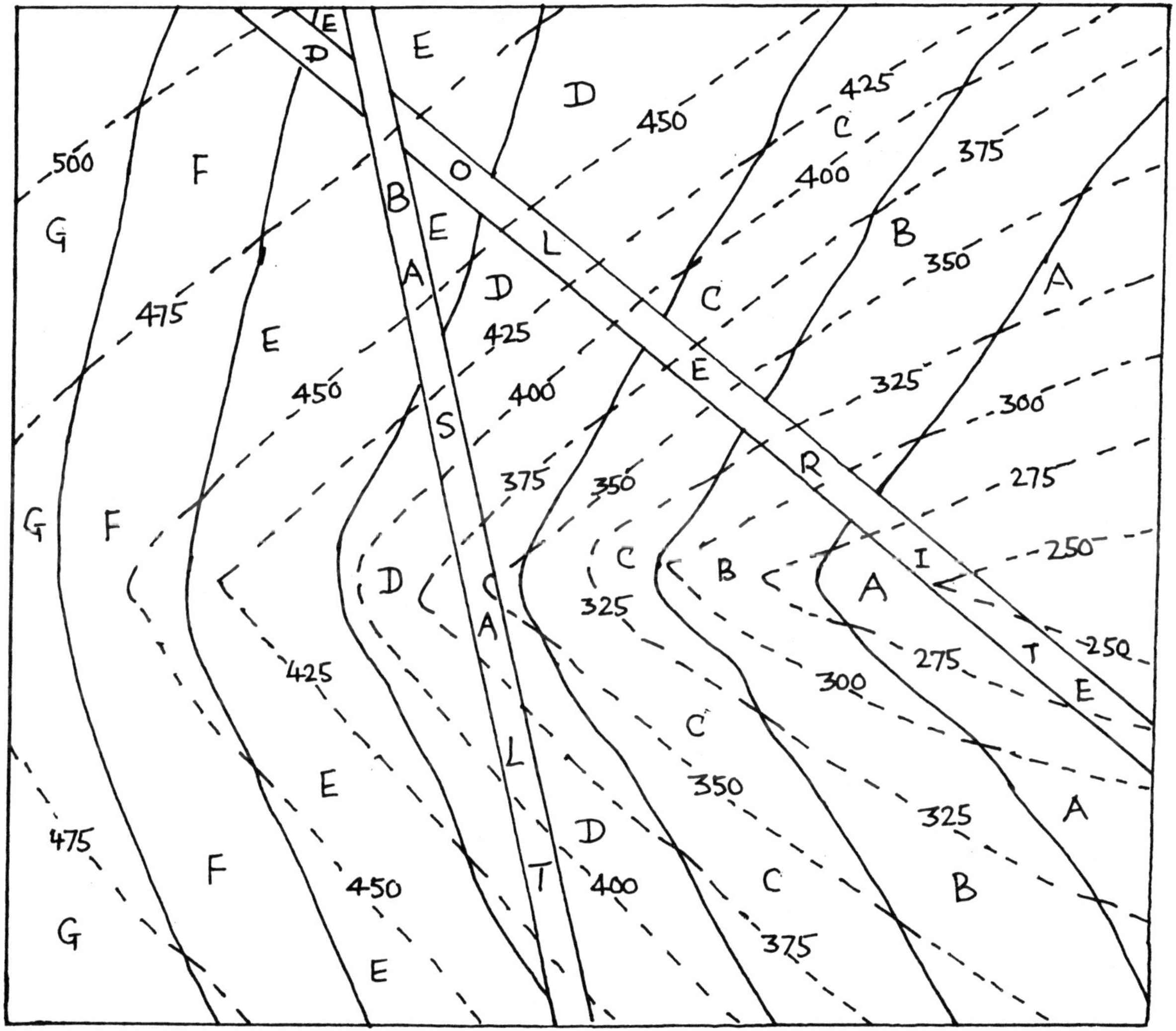

Describe geology of the area. Study the relation between the dykes and the beds, and that between the dykes themselves. Note that the basalt dyke is younger than the dolerite dyke. Draw a suitable geological section.

0 1 cm = 100 ut

Cont. Int. = 25 meters.

Map 17.

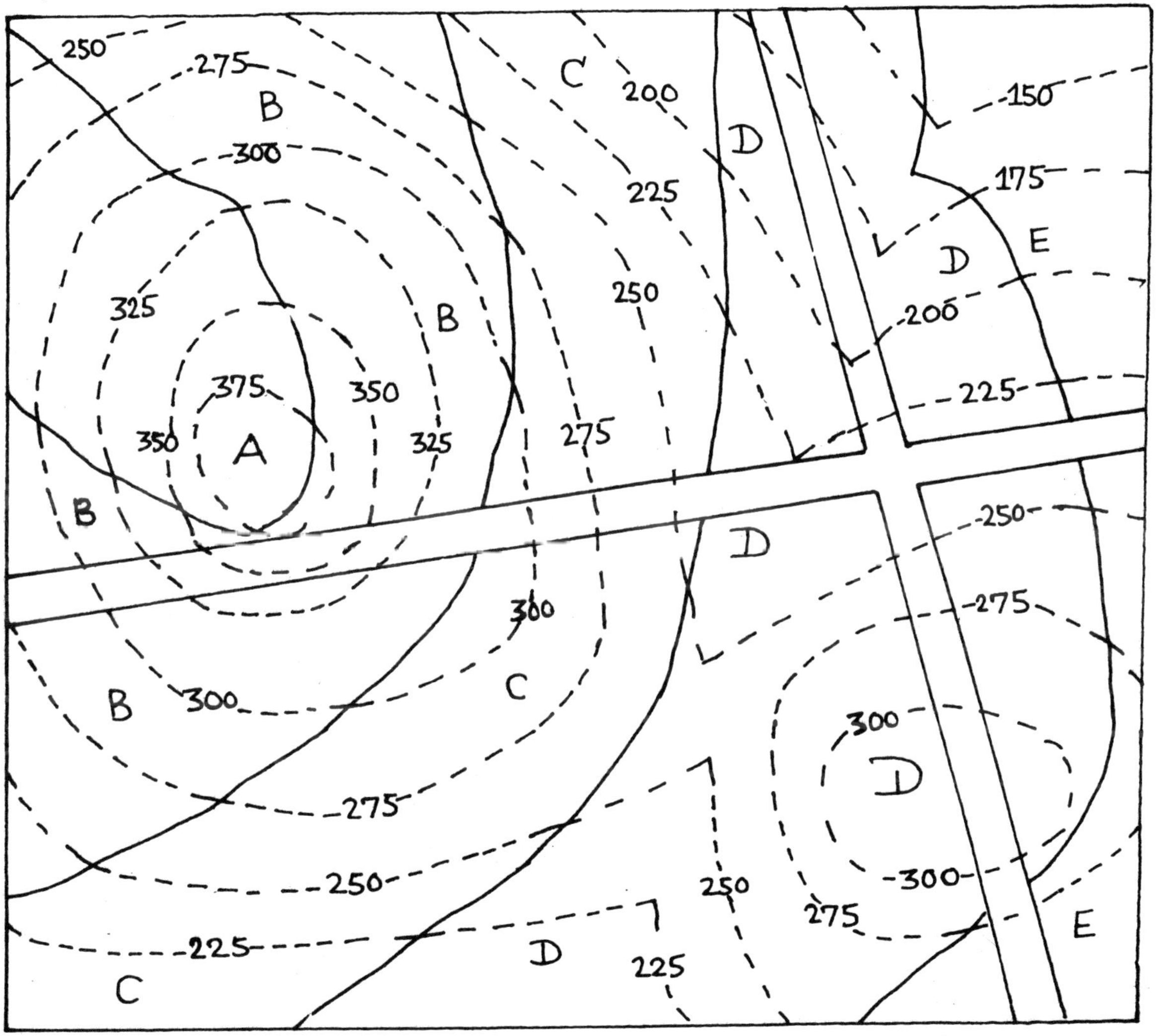

Describe geology of the area. Study the behaviour of the intersecting dykes with each other, and with the beds. Both the dykes are of same age, but are younger than the beds. Draw a suitable geological section.

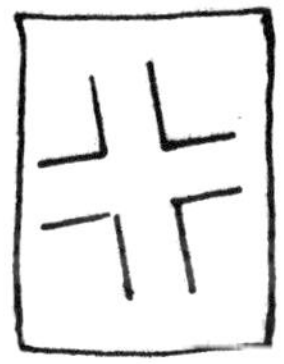

Intersecting dykes.

N

0 1 cm = 100 ut

Cont. Int. = 25 Meters.

Map 18.

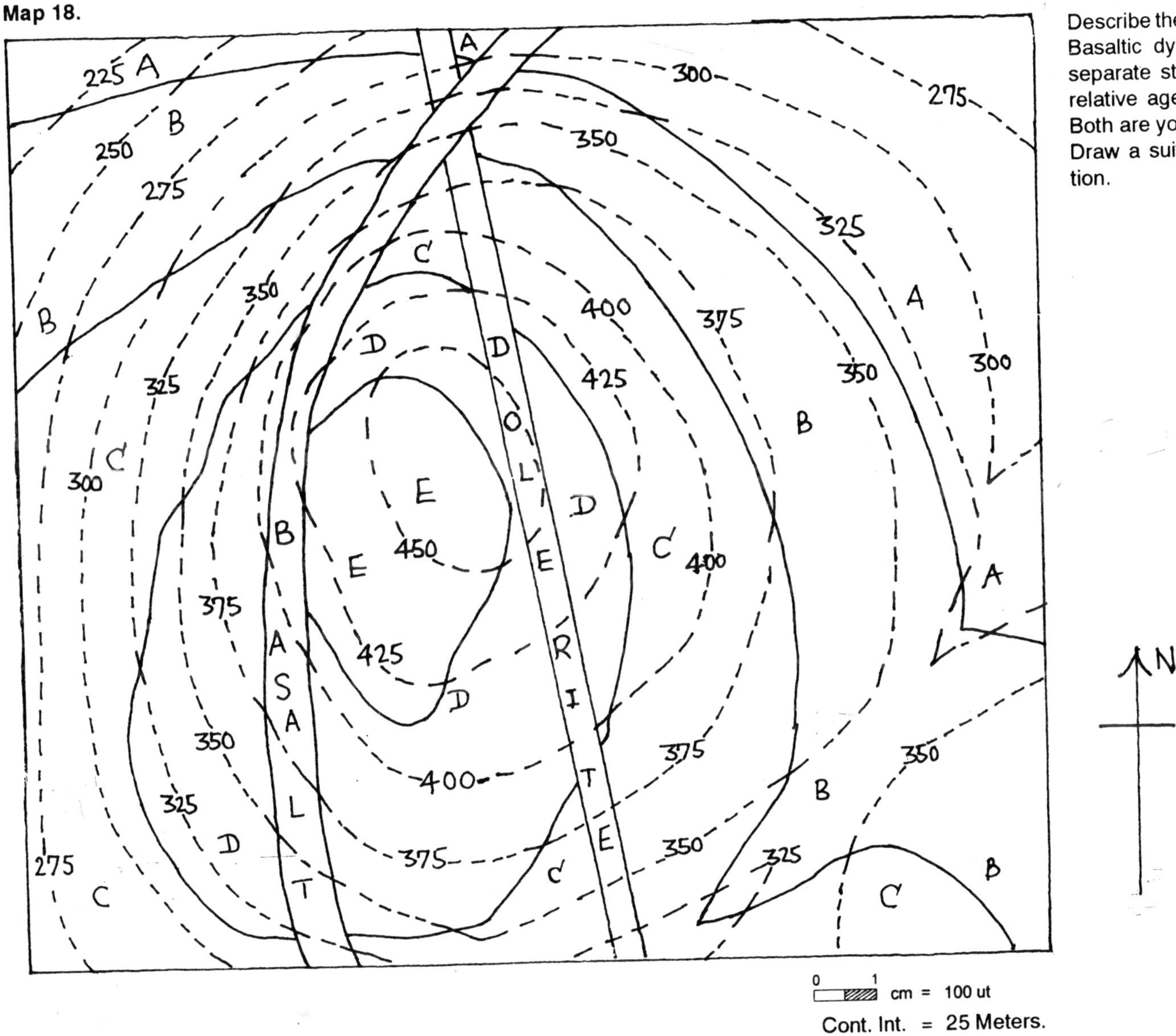

Describe the geology of the area, Basaltic dyke is inclined, draw separate strike lines. Study the relative ages of the two dykes. Both are younger than the beds. Draw a suitable geological section.

Map 19.

Describe geology of the area. Study the beds outcropping across the fault F_1-F_1. Draw a suitable geological section.

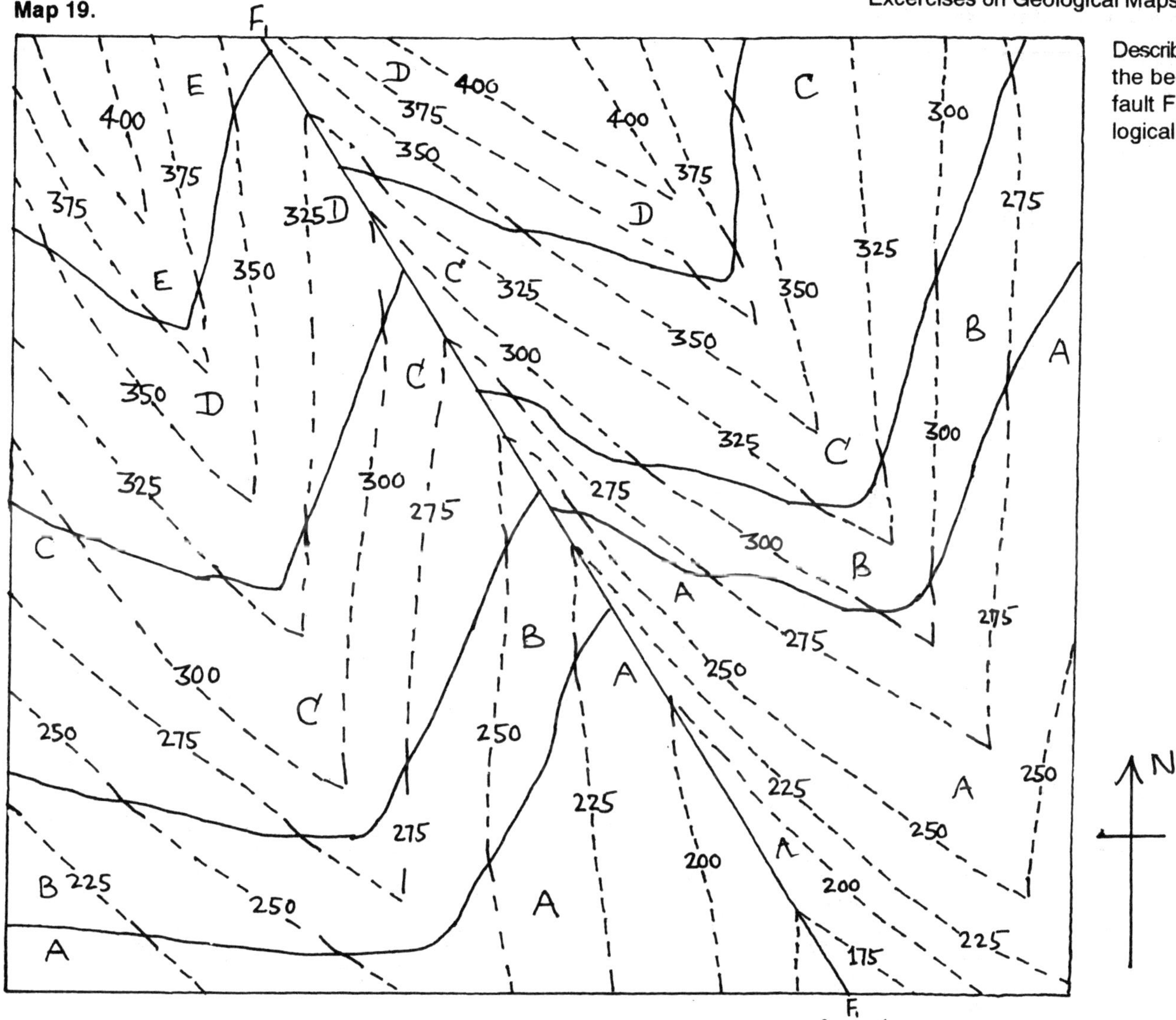

0 1 cm = 100 ut

Cont. Int. = 25 Meters.

Map 20.

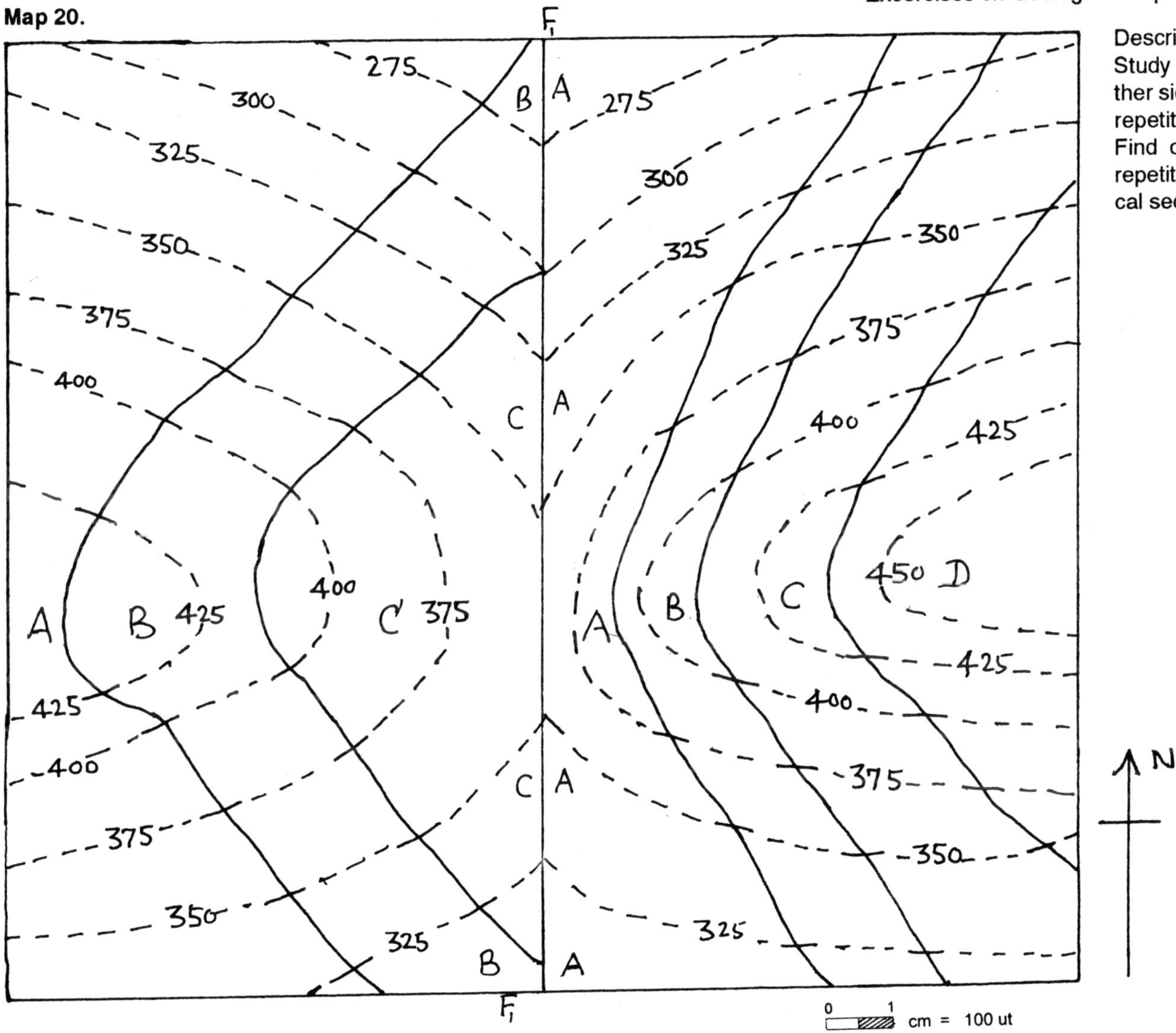

0 1 cm = 100 ut

Con. Int. = 25 Meters

Describe the geology of the area. Study the beds occurring on either sides of fault F_1-F_1. Note the repetition of beds across the fault. Find out the cause for such a repetition. Draw a suitable gelogical section.

Map 21.

Describe geology of the area. Draw a suitable geological section.

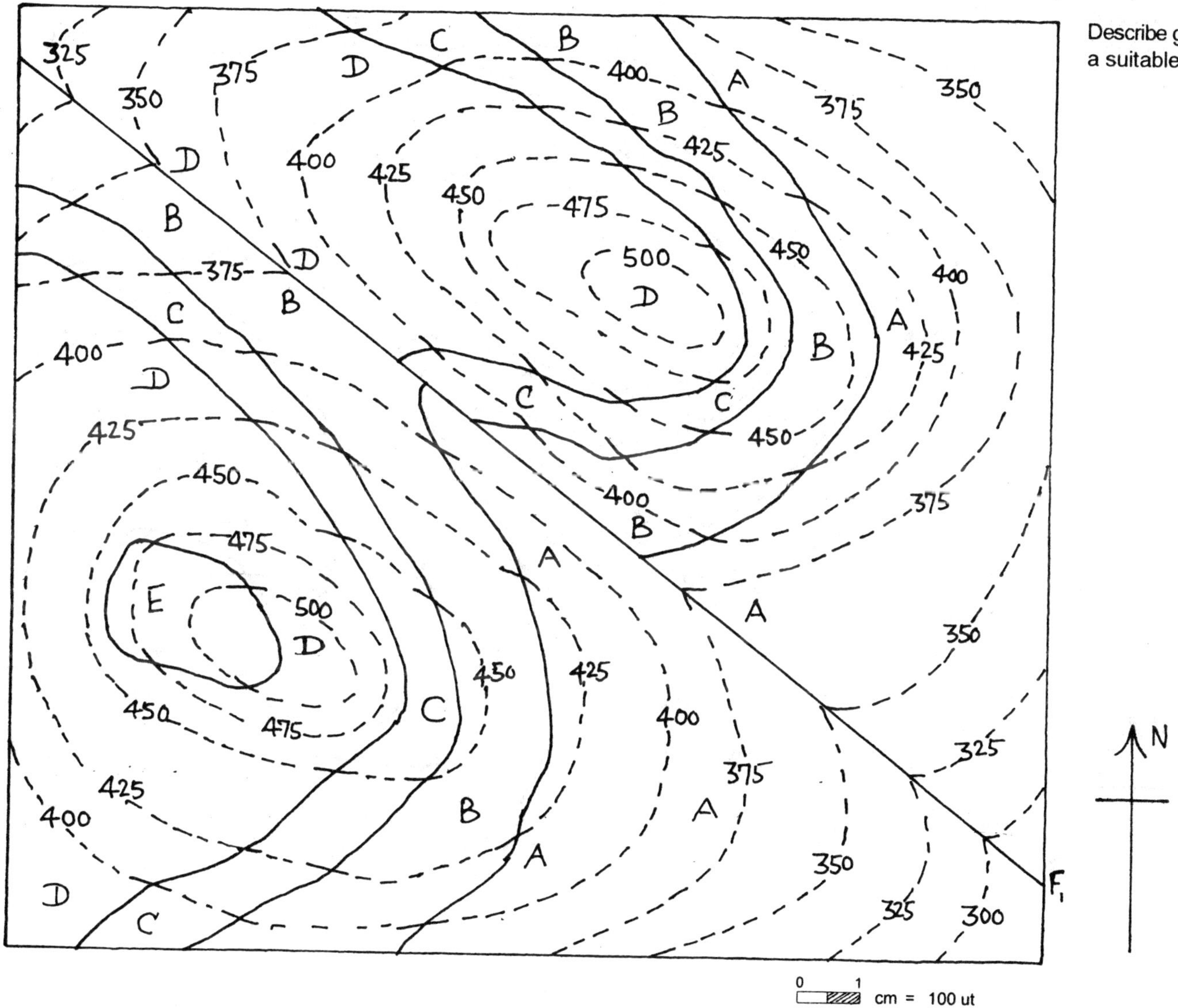

cm = 100 ut

Cont. Int. = 25 Meters.

Map 22.

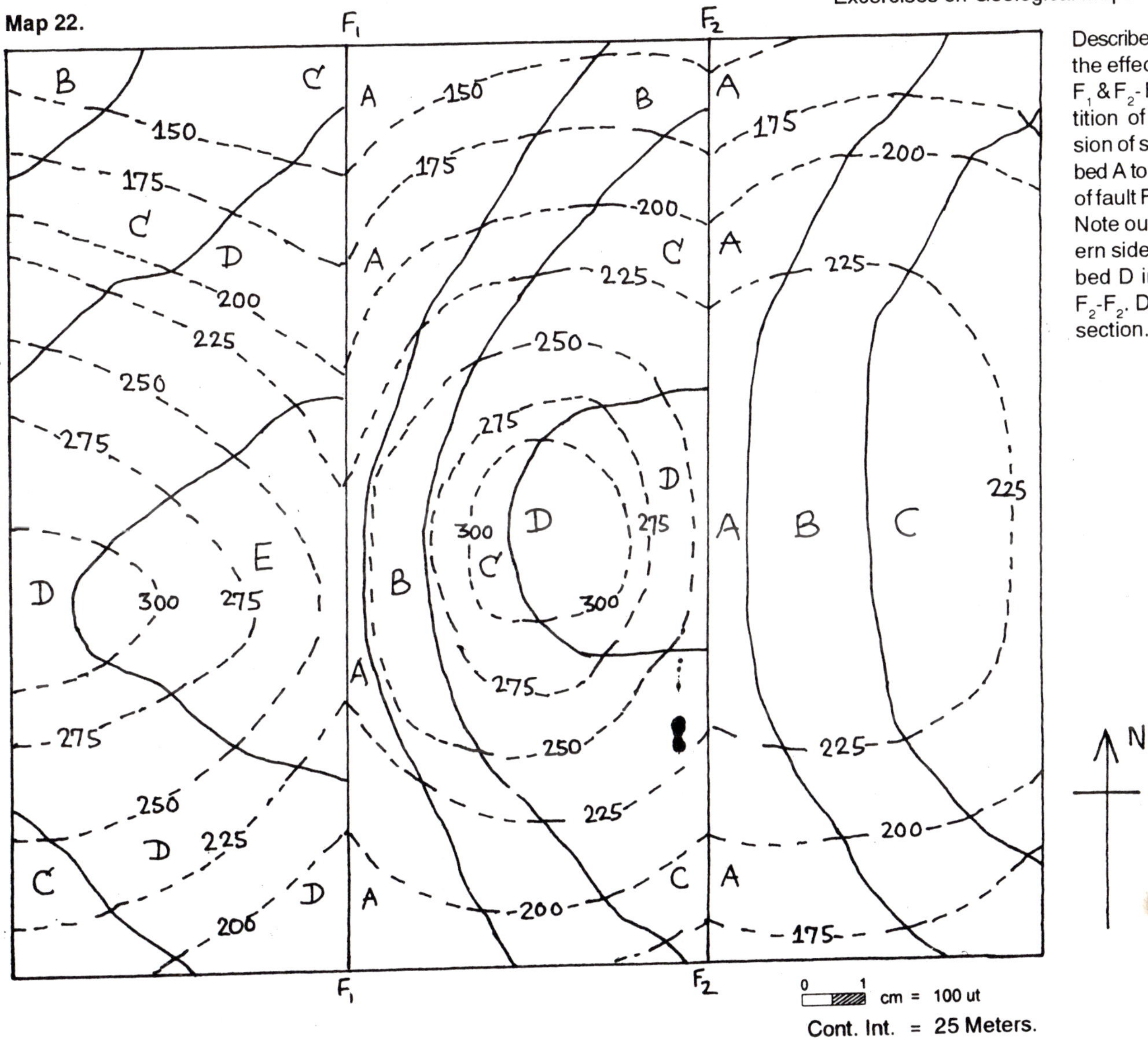

Describe geology of the area. Study the effect of two parallel faults F_1-F_1 & F_2-F_2, separately. Note repetition of some beds while omission of some others. Note inlier of bed A to the immediate easternside of fault F_1-F_1 as well as fault F_2-F_2. Note outlier of bed E on the western side of fault F_1-F_1, and that of bed D in between faults F_1-F_1 & F_2-F_2. Draw a suitable geological section.

Map 23.

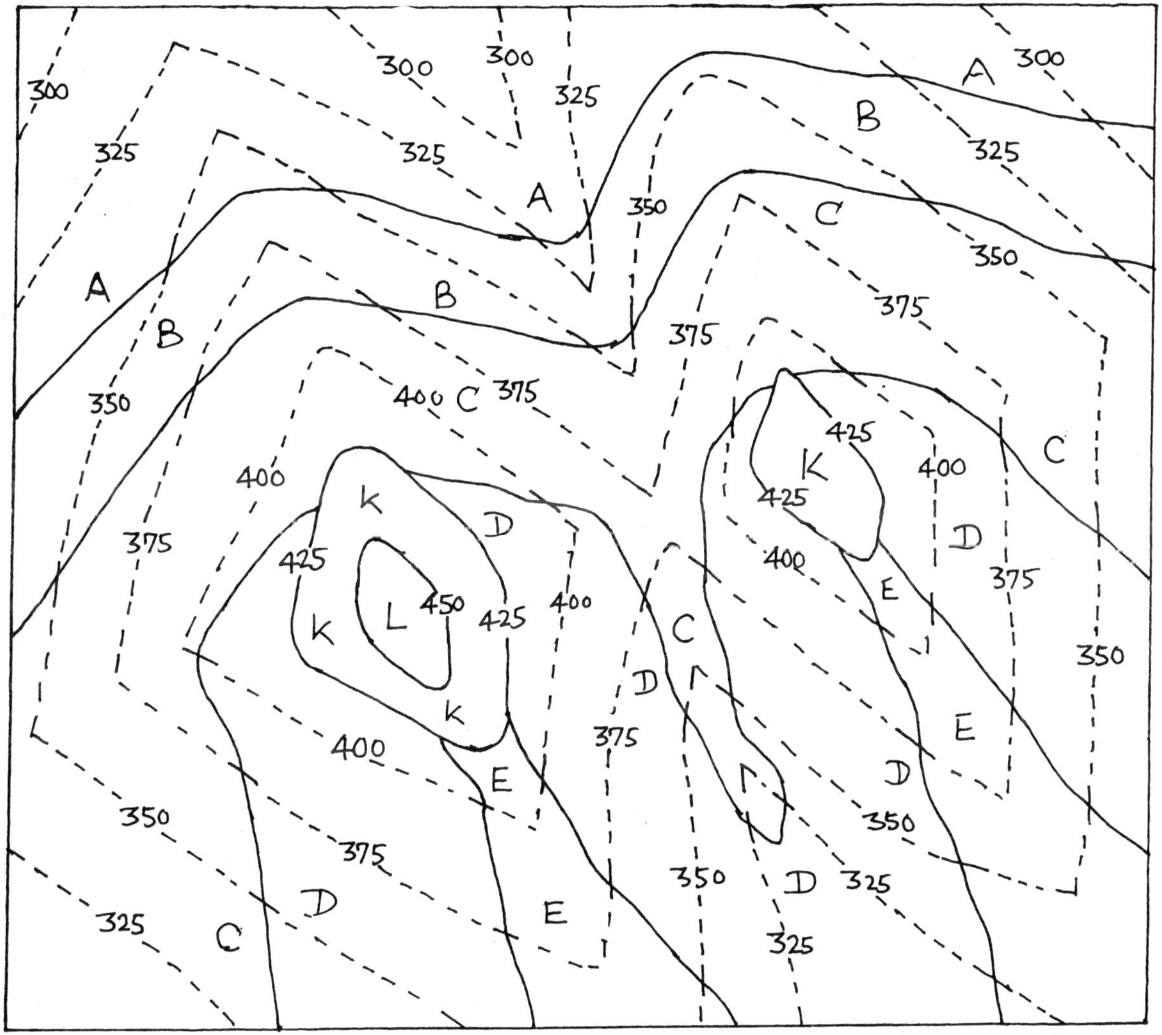

0 1 cm = 100 ut

Cont. Int. = 25 Meters.

Describe geology of the area. Note that the bottom of K bed marks the surface of unconformity. Comment on the repetition of K bed. Draw a suitable geological section.

Map 24.

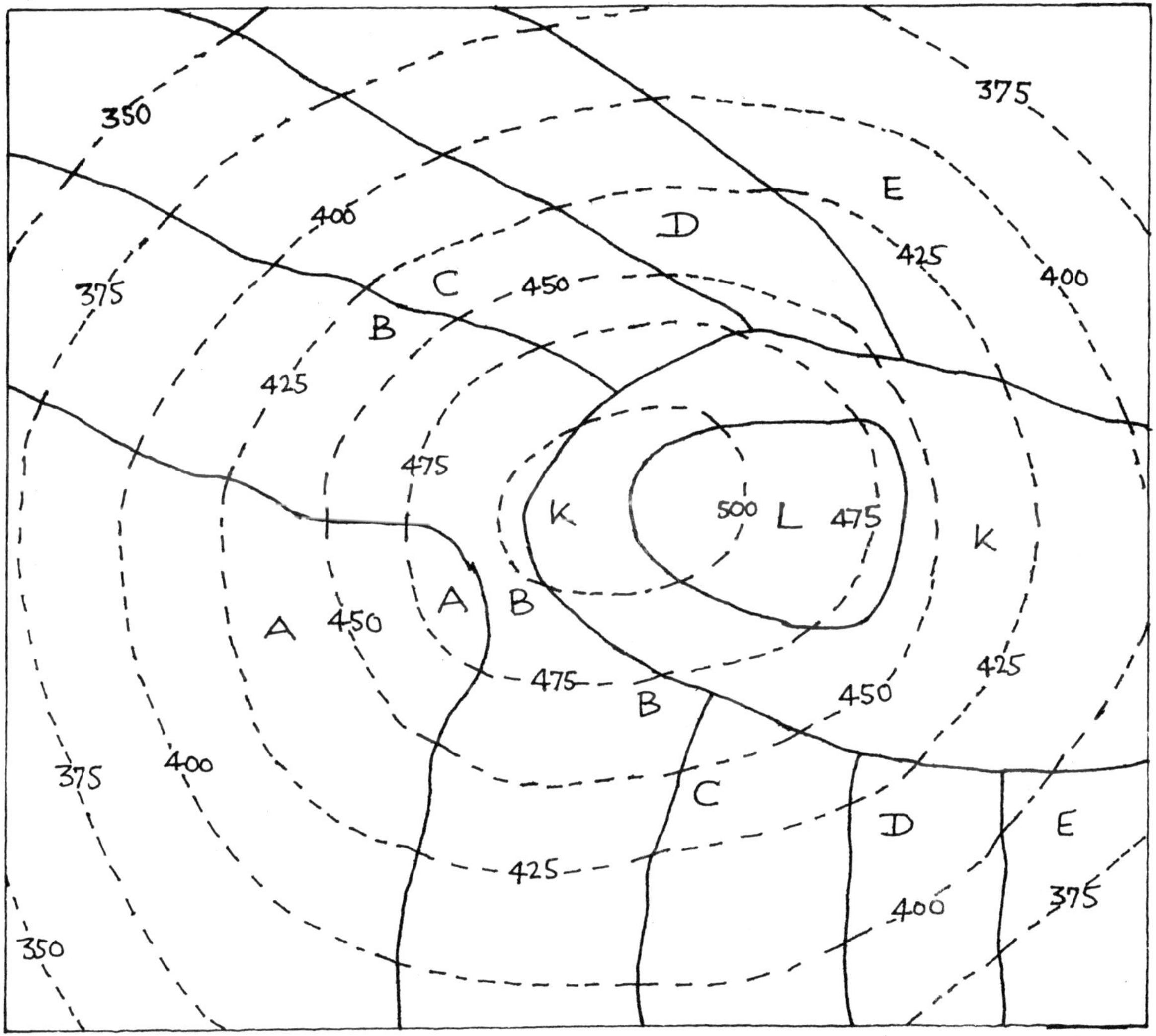

Describe the geology of the area. Note that both the series of beds are dipping and have produced a perfect angular unconformity. Bed L occurs as an outlier. Draw a suitable geological section.

0 1 cm = 100 ut

Cont. Int. = 25 Meters.

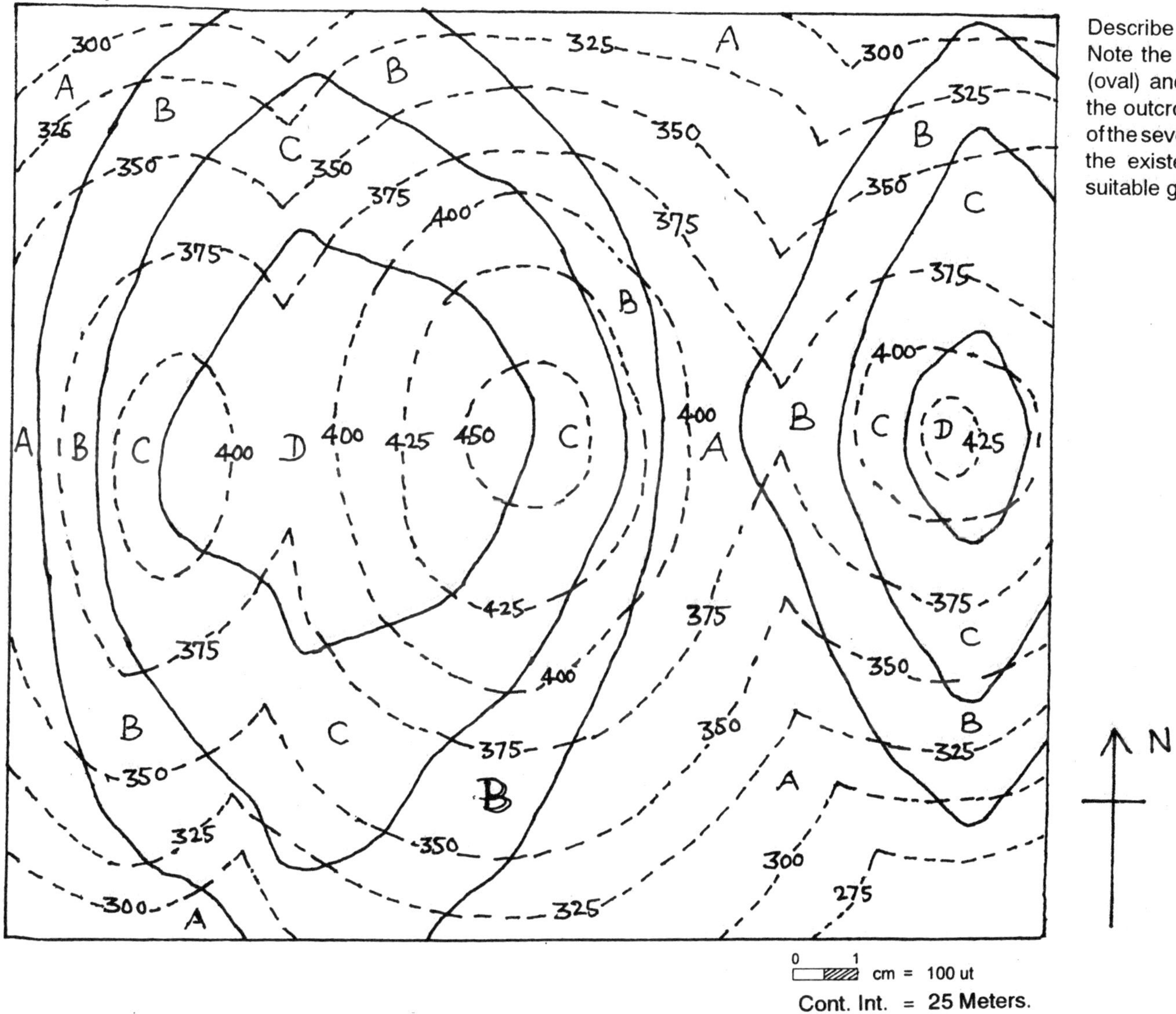

Describe the geology of the area. Note the characteristic diamond (oval) and hour-glass shapes of the outcrops. Note the repetition of the several beds which is due to the existence of folds. Draw a suitable geological section.

Map 26.

Describe the geology of the area. Note the asymmetric nature of the fold, and the reversal of topography. Draw a suitable geological section.

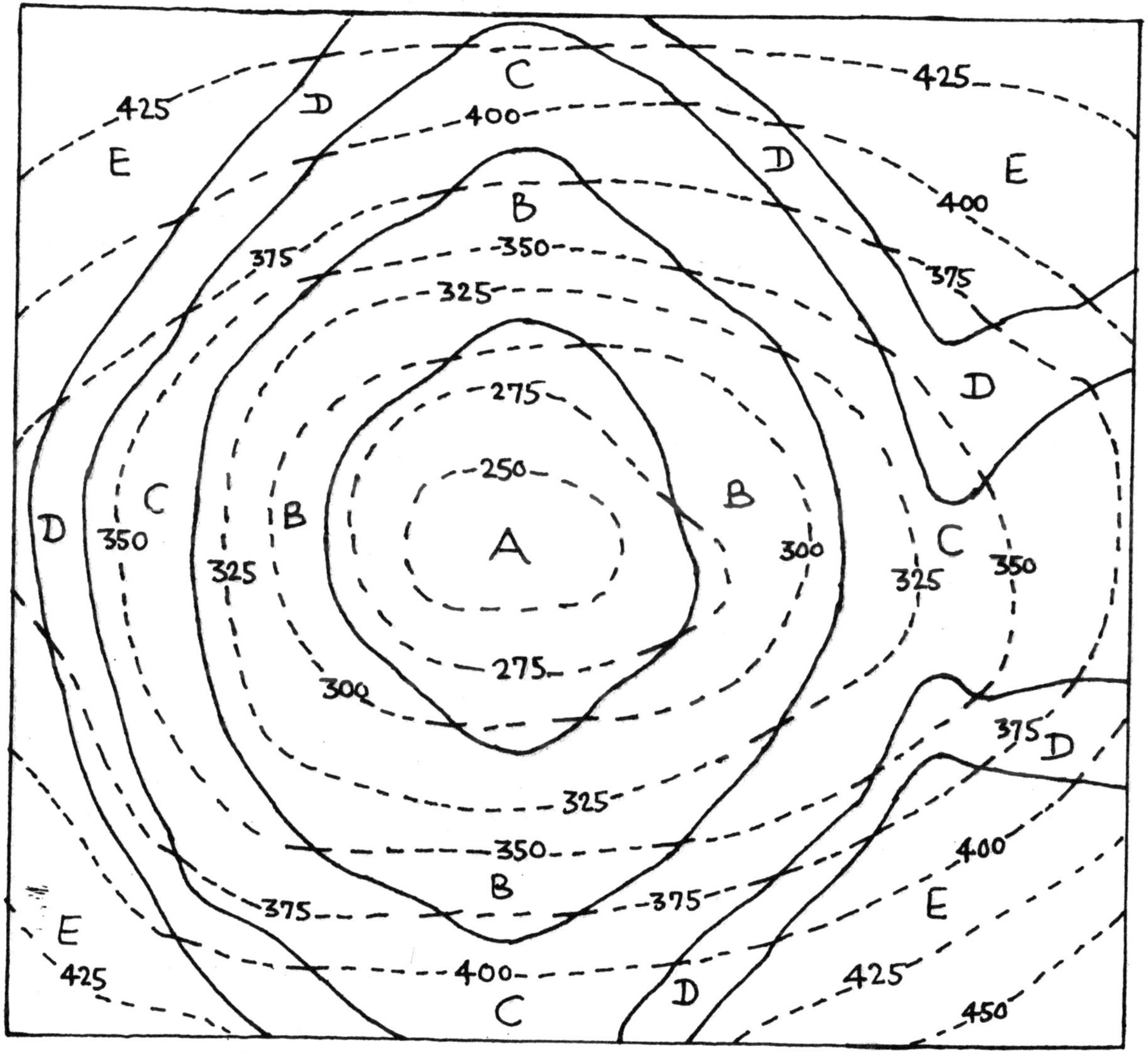

0 1 cm = 100 ut

Cont. Int. = 25 Meters.

Map 27.

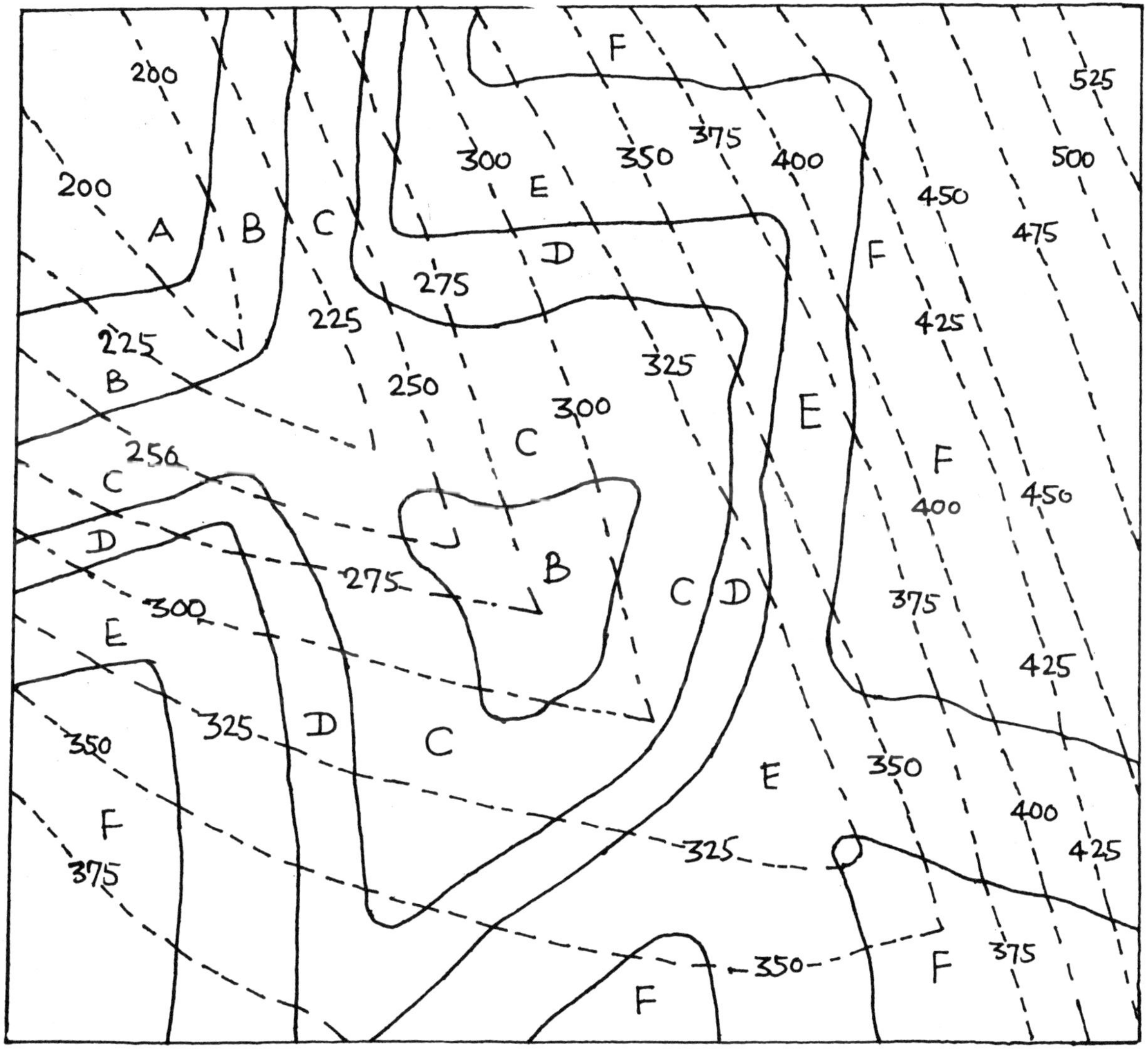

Describe geology of the area. Study the shapes of the outcrops carefully. Count the number of synclines and anticlines. Note the development of inlier of bed A towards NW corner of the area, and of bed B in the central part of the area. Draw a suitable geological section.

0 1 cm = 100 ut

Cont.Int. = 25 Meters.

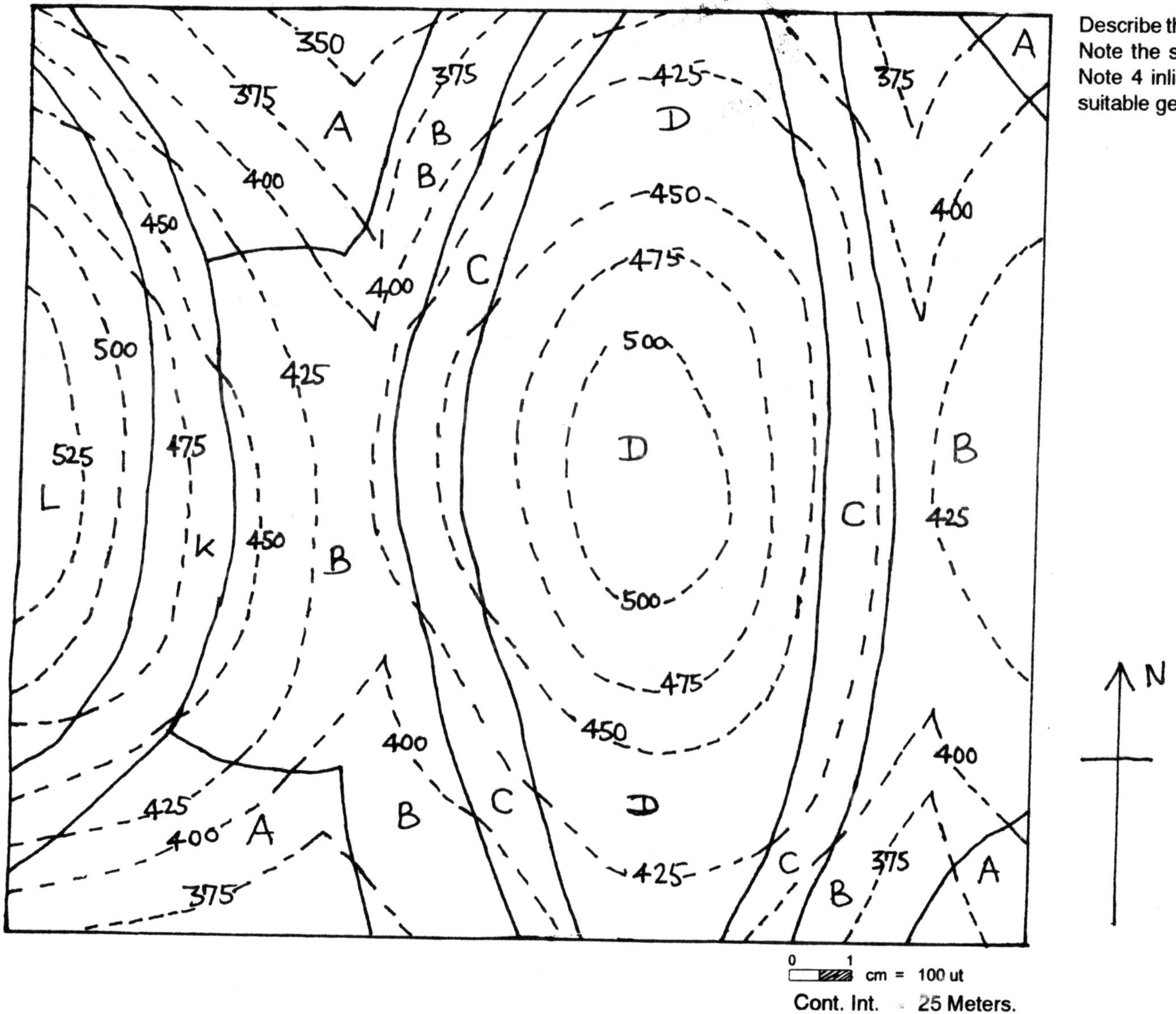

Describe the geology of the area. Note the shape of the syncline. Note 4 inliers of bed A. Draw a suitable geological section.

Map 29.

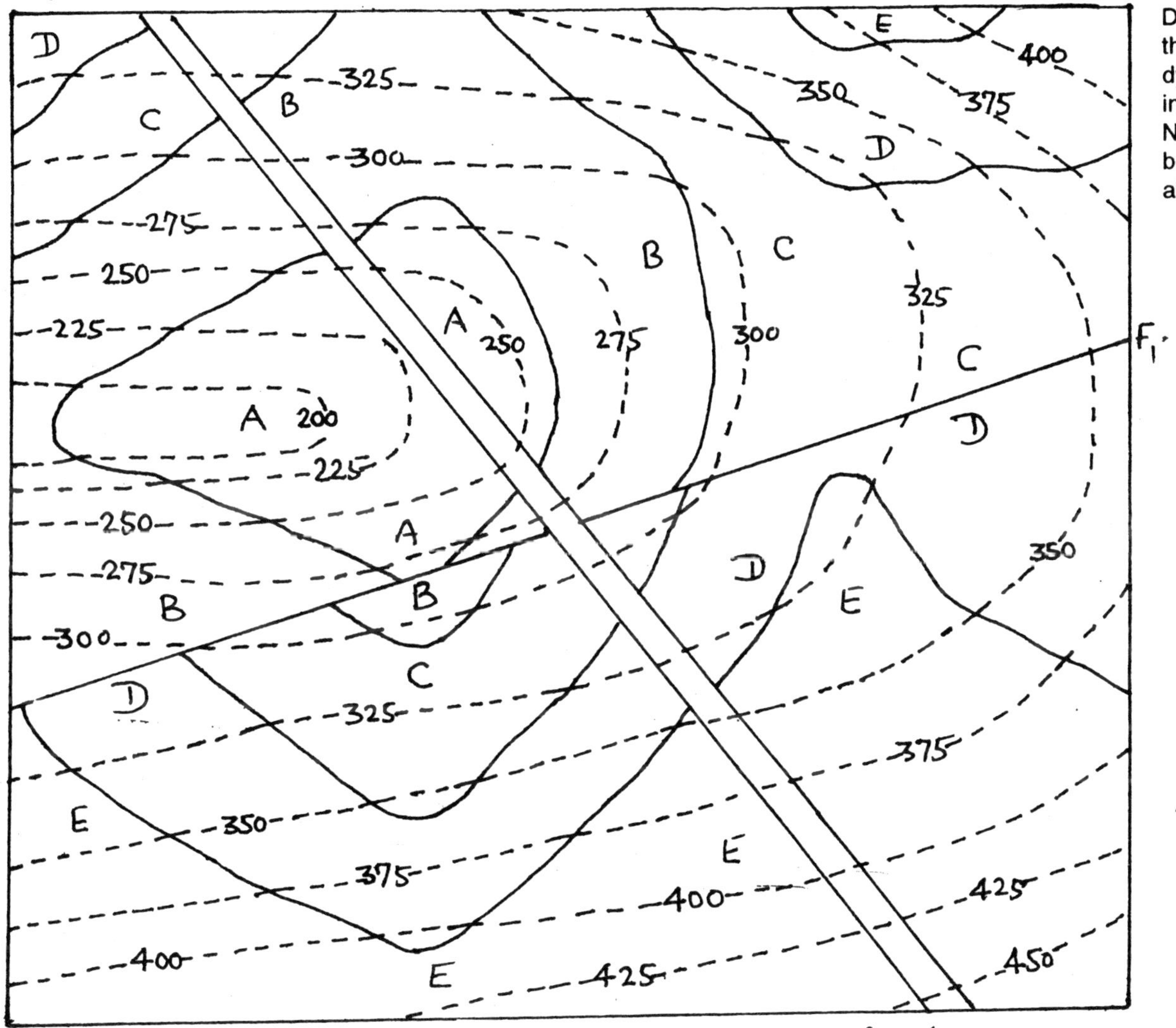

0 1 cm = 100 ut

Cont. Int. = 25 meters.

Describe geology of the area. Study the nature of contact between the dyke and the fault. Note the shift in the outcrops along the fault. Note the development of inlier of bed A in the valley. Draw a suitable geological section.

Map 30.

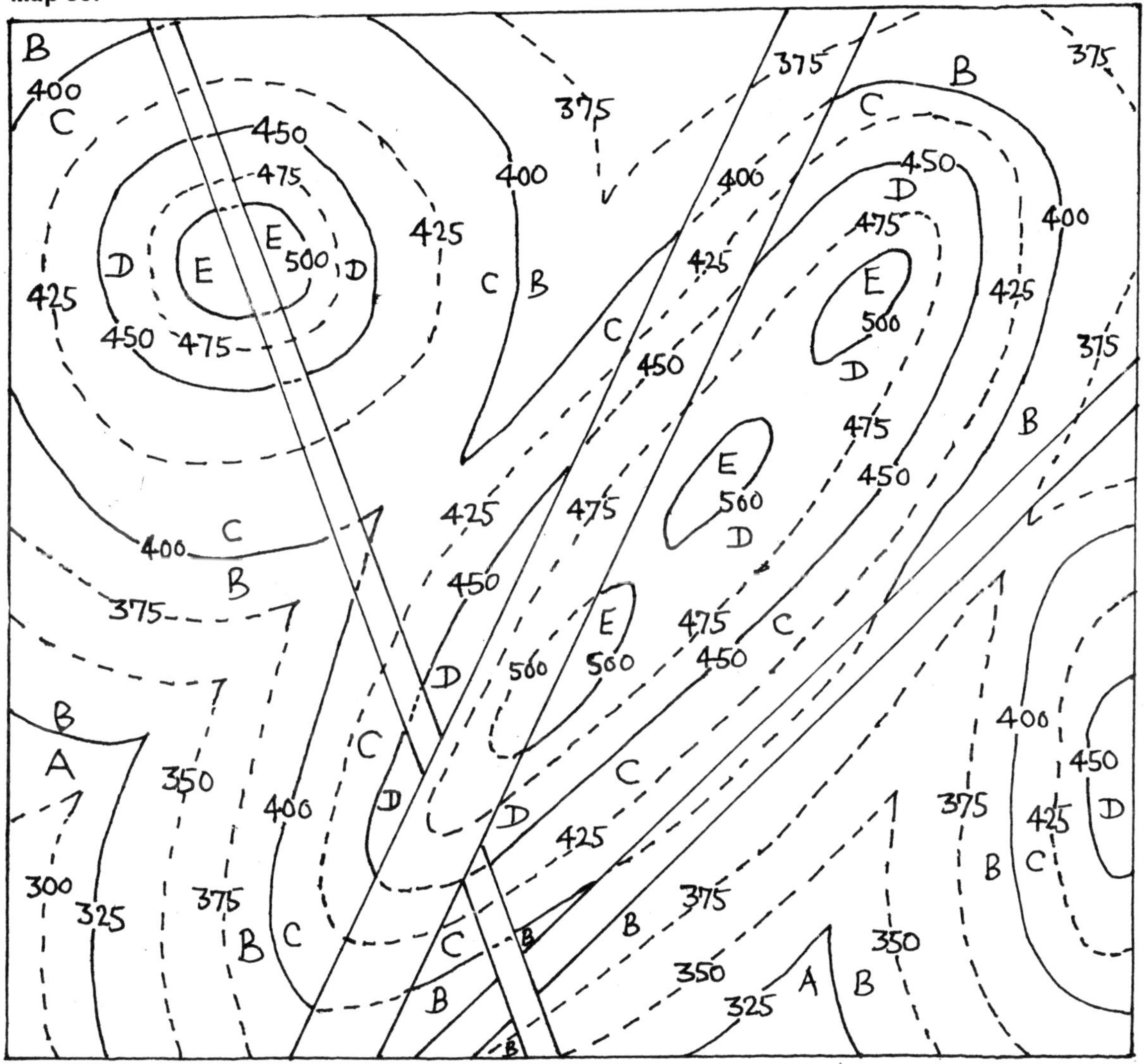

Describe the geology of the area. Study the relation between the three intersecting dykes, and determine their relative ages. Note that there are 4 outliers of bed E and 1 of bed D (eastern border of the area). Contours 325, 400, 450 and 500 also mark bedding planes, and are therefore shown as continuous lines, except where these cross into the dyke. Draw a suitable geological section.

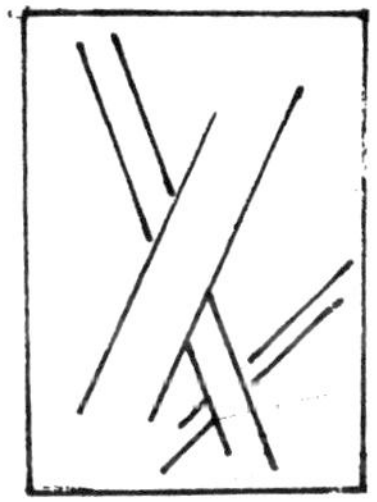

Three intersecting dykes.

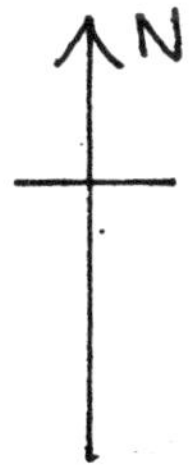

0 1 cm = 100 ut

Cont. Int = 25 Meters.

Additional Exercises on Dipping Beds

A. Define terms dip, true dip, apparent dip, true strike and apparent strike. Which is more, true dip or apparent dip?

B. Describe how the gradient of a dipping bed is calculated from a geological map. Which gradient (dip) is steep; 1 in 10 or 1 in 4? How is it so?

C. Define terms inlier, outlier and angular unconformity.

D. Describe the procedure of establishing the order of superposition in the case of horizontal and dipping beds, respectively.

E. Describe the procedure of determining the throw of a fault.

F. Describe the difference between appearance of horizontal beds and the dipping beds on a contour map.

G. Describe the procedure of establishing the relative ages of intersecting dykes.

H. Describe the normal shape of an anticline and a syncline.

I. Schematically show the relation between the strike lines and the trends of the dip, the strike and the oblique faults.

SECTION--IV

TRACING OF OUTCROPS, i.e., CONSTRUCTION OF GEOLOGICAL MAPS

HORIZONTAL BEDS
INCLINED BEDS

HORIZONTAL BEDS

Map 31. Initial outcrop point given on contour by way of a cross.

Map 32. Initial outcrop point stated in the text.

INCLINED BEDS

Map 33. Data given in the form of a continuous line. Note that two consecutive basic strike lines are readily obtainable.

Map 34. Data given in the form of a continuous line on two consecutive contour lines. Note that two basic strike lines are immediately available from which the amount of dip and direction can be calculated.

Map 35. Data given in the form of a continuous line but not on two consecutive contour lines. Note that a strike line is required to be introduced in between the two strike lines initially drawn. Only then the correct amount of dip can be obtained.

Map 36. Data given in the form of true dip amount and direction, and true strike in the text, for a point marked by +/- on one contour. This gives only one strike line, others are to be drawn considering the given amount of dip.

Map 37. Data given in the text for two points marked by +/- on one contour, in the form of dip amount and general direction of dip. This gives only one basic strike line. Others are to be drawn considering the given amount of dip.

Map 38. Four places marked by +/- on same contour are given. These points give the first basic strike line. Others are to be drawn with the help of amount of dip and its direction, as given in the text.

Map 31.

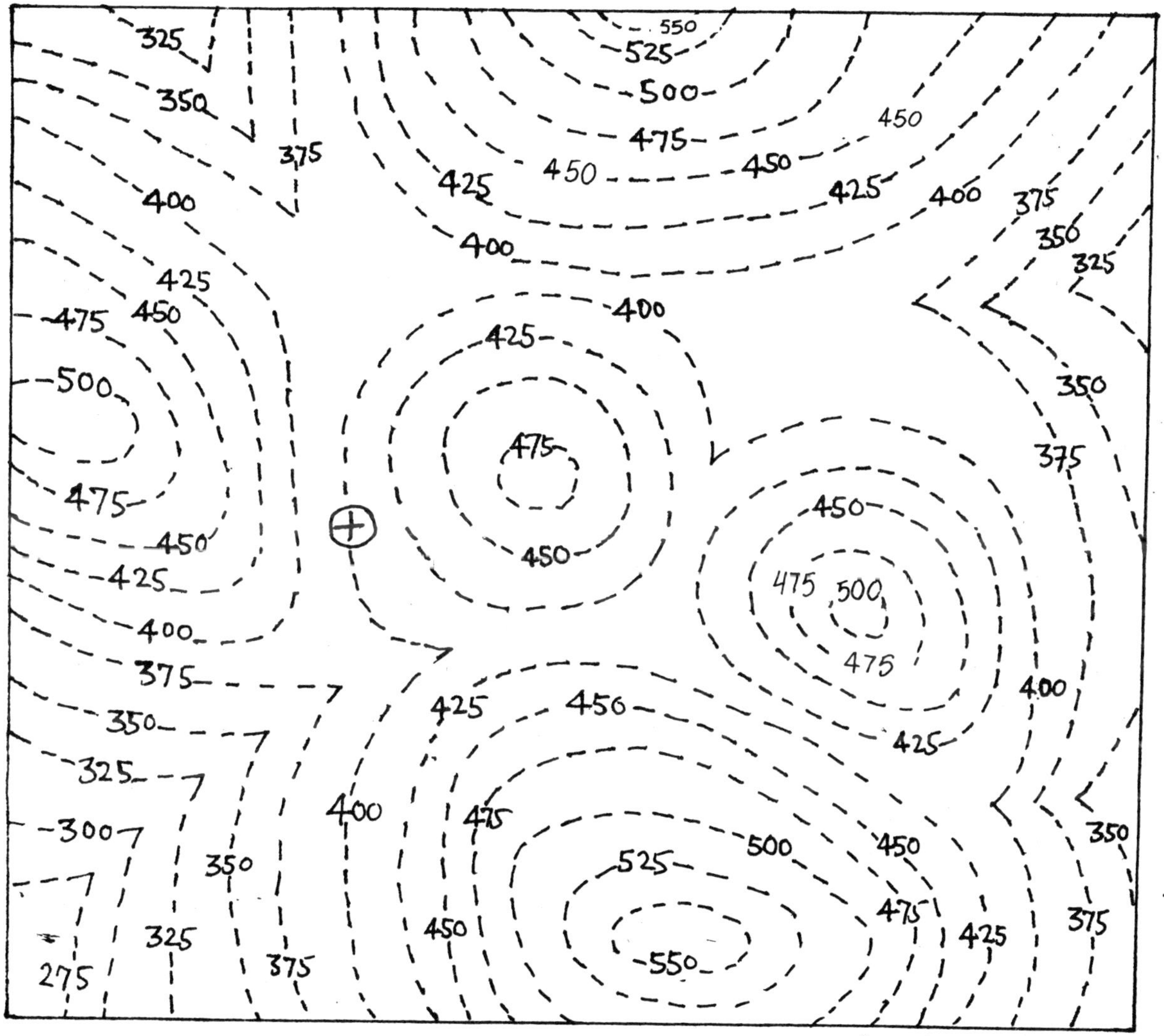

0 1 cm = 100 ut

Cont. Int. = 25 Meters.

At a place marked +/- the bottom of bed c crops out. The other beds of the conformable horizontal series, their thicknesses and the order of super position, is given below.

F	
E	50 Meters
D	75 Meters
C	25 Meters
B	50 Meters
A	

Trace the outcrops of the several beds.

Map 32.

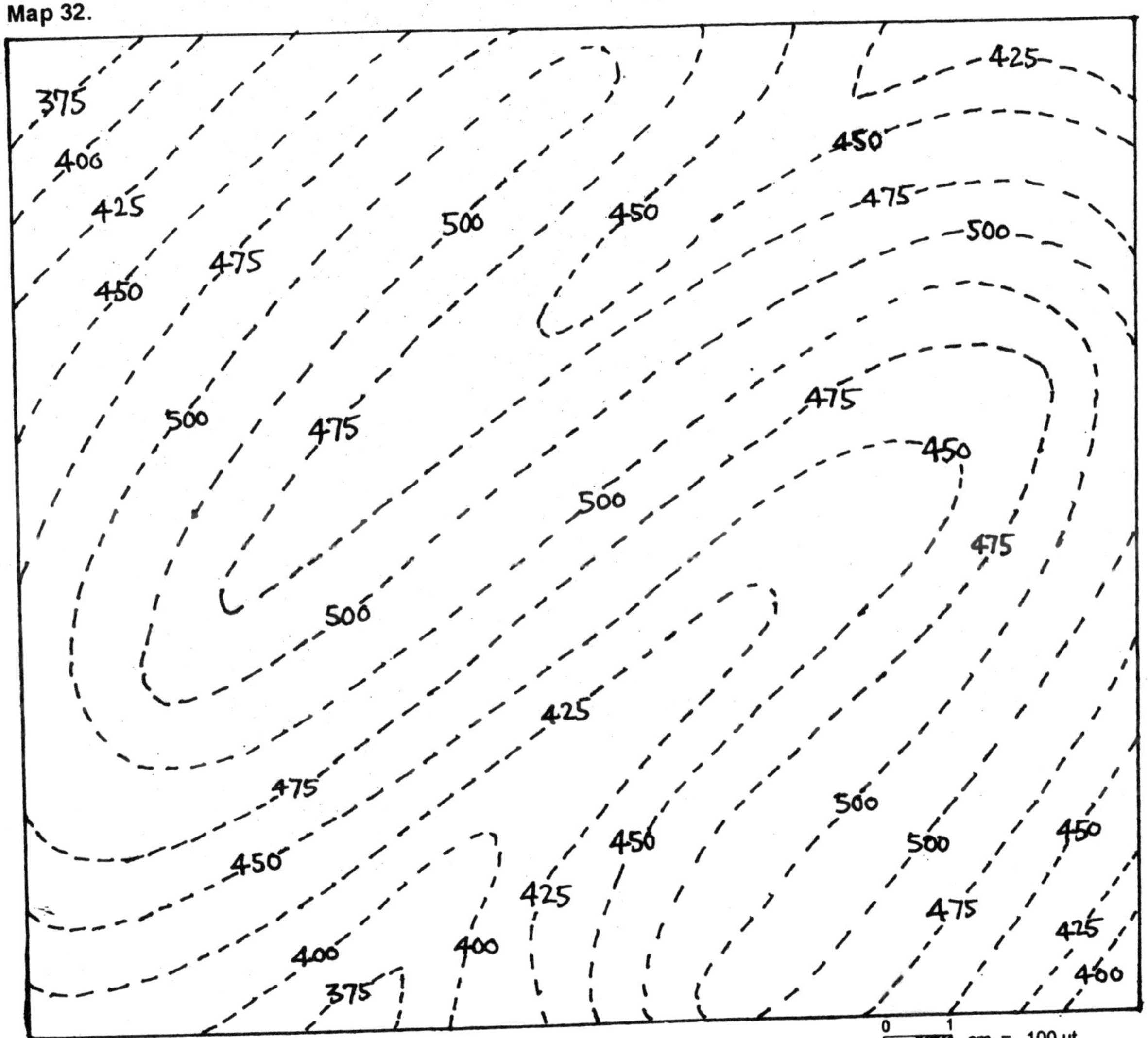

0 1 cm = 100 ut

Cont. Int. = 25 Meters.

Top of bed A crops out at a height of 375 meters. The conformable series consists of horizontal beds. Their thicknesses and the order of superposition, are given below.

E	
D	75 meters
C	25 meters
B	25 meters
A	

Trace the outcrops of the several beds.

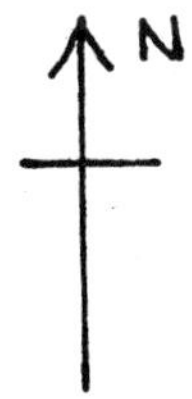

Map 33.

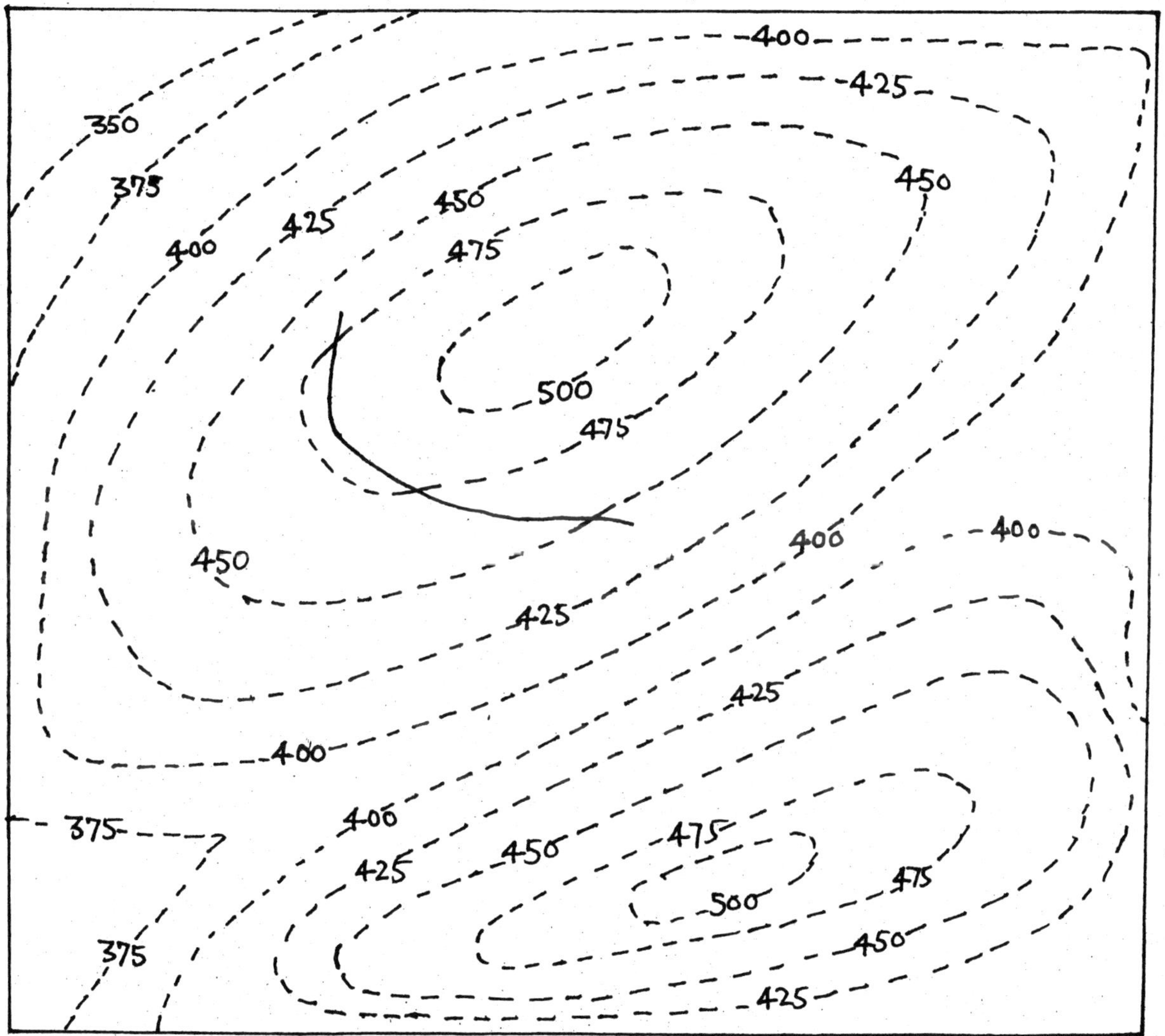

In the area is exposed a conformable series of dipping beds. The continuous line marks the top of bed C. The thicknesses and the order of superposition of the several beds, is as given below.

E	
D	50 Meters
C	25 Meters
B	75 Meters
A	

Trace the outcrops of the several beds. Calculate the amount of true dip and its direction. Also find the direction of true strike.

0 1 cm = 100 ut eters.

Cont. Int. = 25 meters.

Map 34.

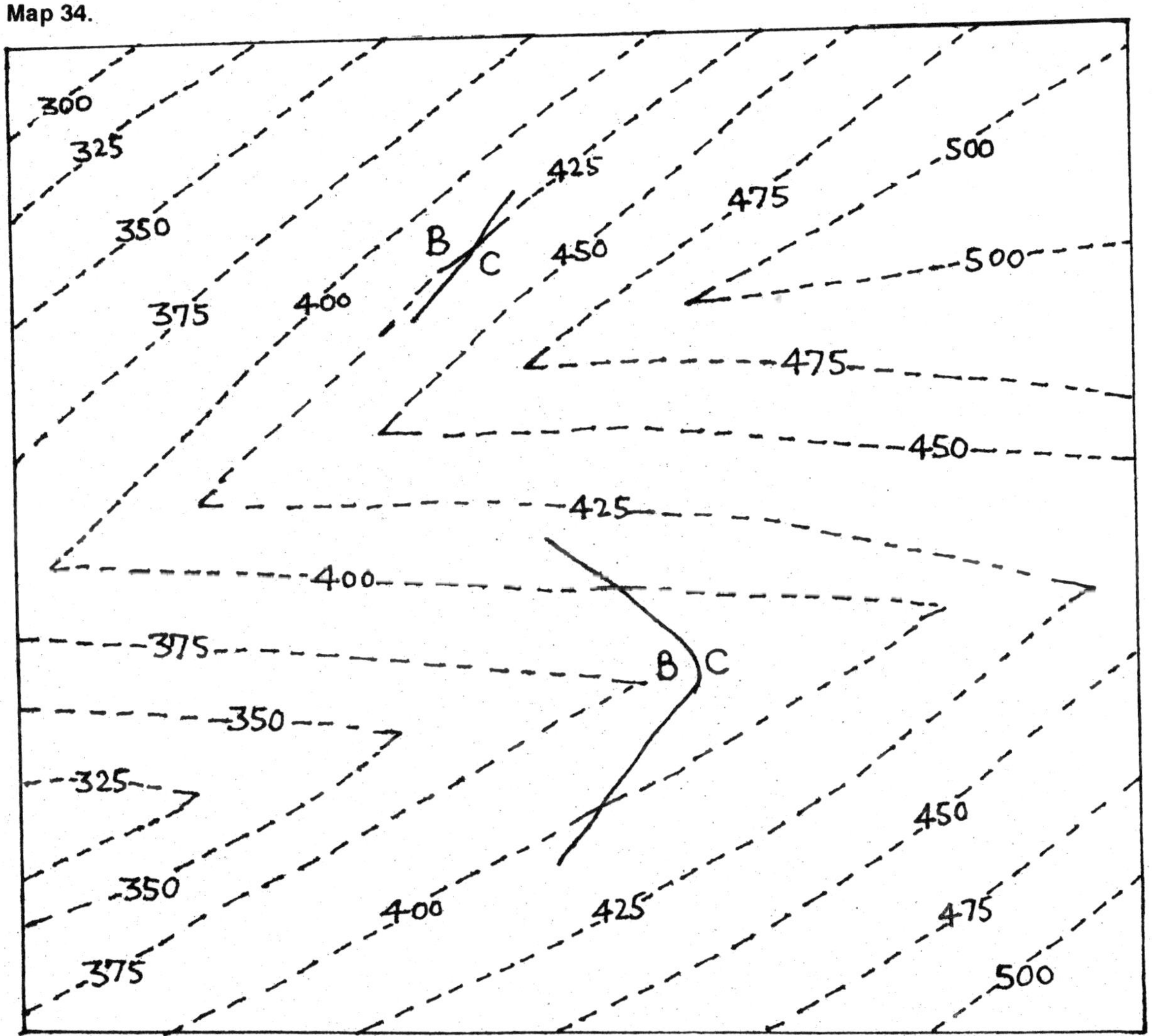

0 1 cm = 100 ut ters.

Cont. Int. = 25 Meters.

Continuous line marks the B/C bedding plane. The thicknesses and the order of superposition of a conformable dipping series of beds, is given below.

F	
E	75 Meters.
D	25 Meters
C	50 Meters
B	75 Meters
A	

Trace the outcrops of the several beds. Determine the direction of true dip and true strike, and the amount of true dip.

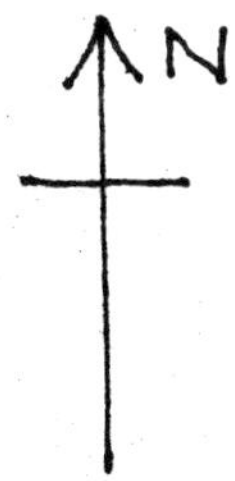

Map 35.

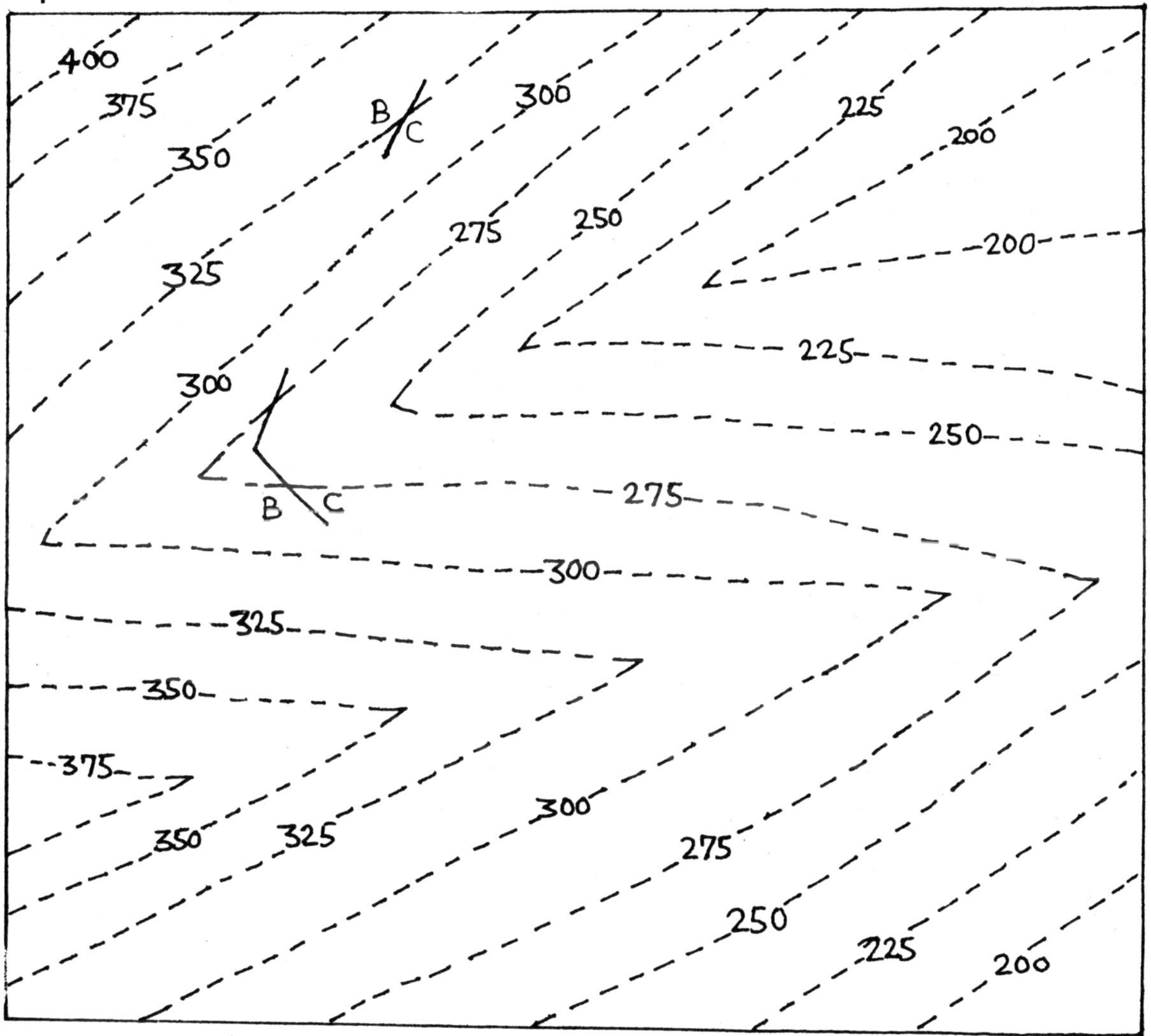

0 1 cm = 100 ut eters.

Cont. Int. = 25 Meters

The continuous line marks the bedding plane B/C of a conformable dipping series. The thicknesses and the order of superposition of the beds, is given below.

A	
B	100 meters
C	75 Meters
D	25 Meters
E	50 Meters
F	

Trace the outcrops of the several beds. Determine the direction of true dip and true strike, and the amount of true dip.

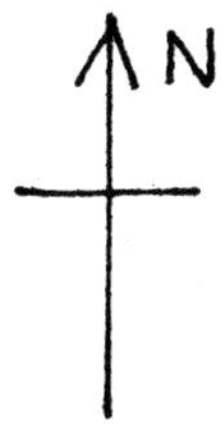

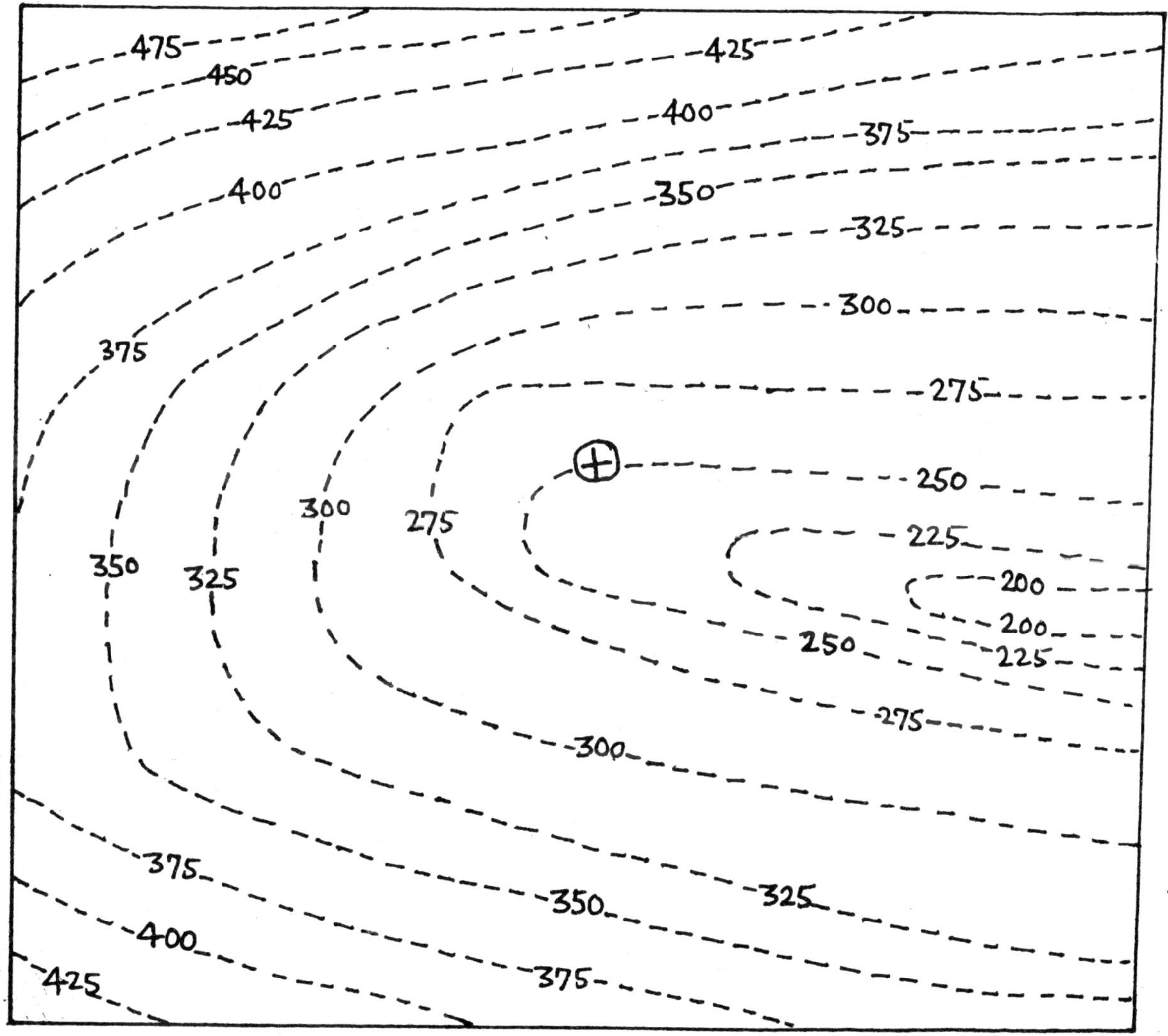

At a place marked +/-, top of bed B crops out, which dips 1 in 6 due 583° E, the strike direction being N7°E-$7°W. The thicknesses and the order of superposition of the beds, is given below.

Bed	Thickness
F	
E	75 Meters
D	100 Meters
C	100 Meters
B	75 Meters
A	

Trace the outcrops of the several beds.

0 1 cm = 100 ut eters.

Cont. Int. = 25 meters.

Map 37.

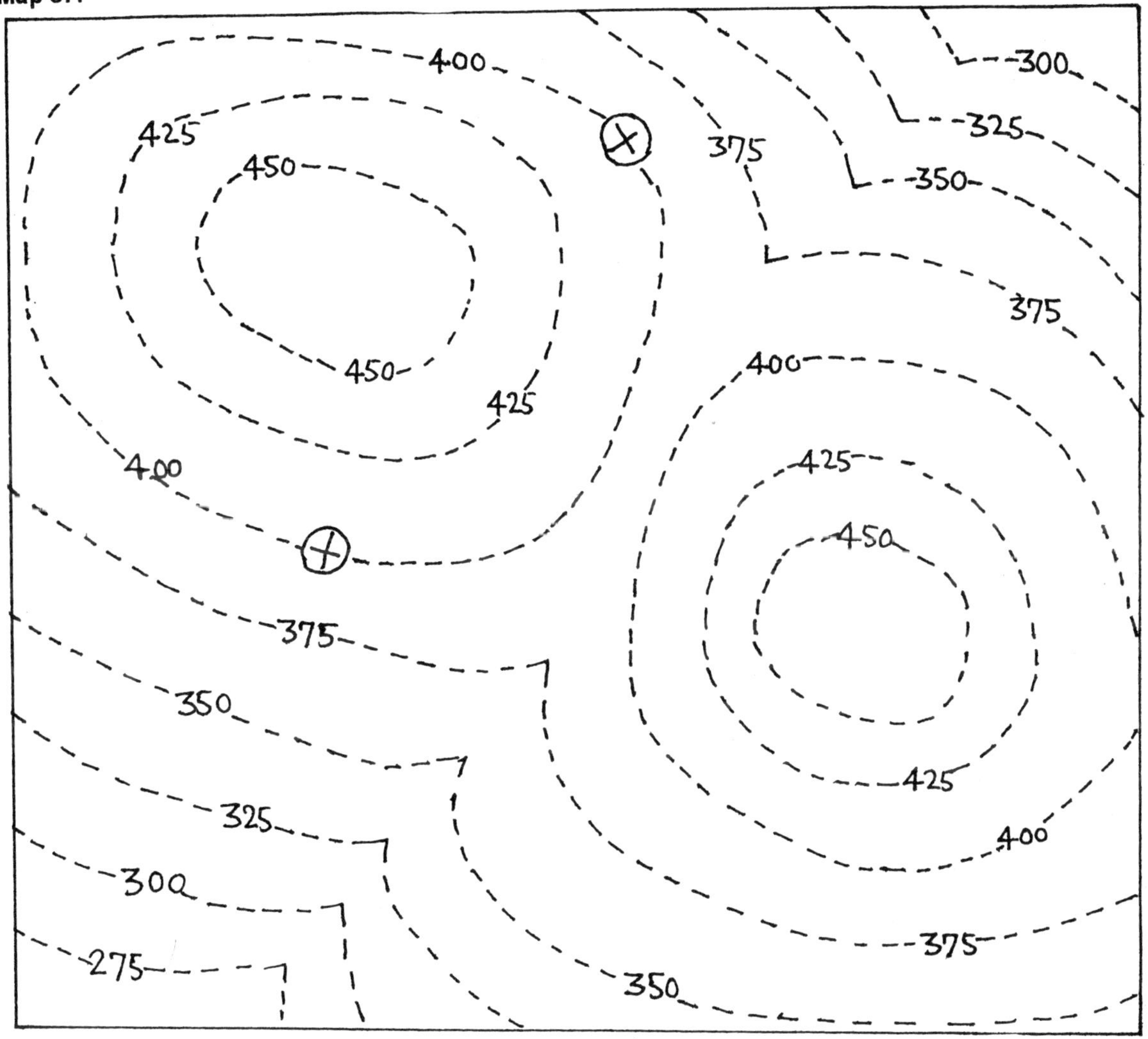

C/D bedding plane crops out at places marked +/-. The conformable dipping series of beds dips 1 in 8 towards general northwest direction. The thicknesses and the order of superposition, is given below.

E	
D	75 Meters
C	75 Meters
B	50 Meters
A	

Trace the outcrops of the several beds. Determine the exact direction of dip and strike of the series.

0 1 cm = 100 ut ers.

Cont. Int. = 25 Meters.

Map 38.

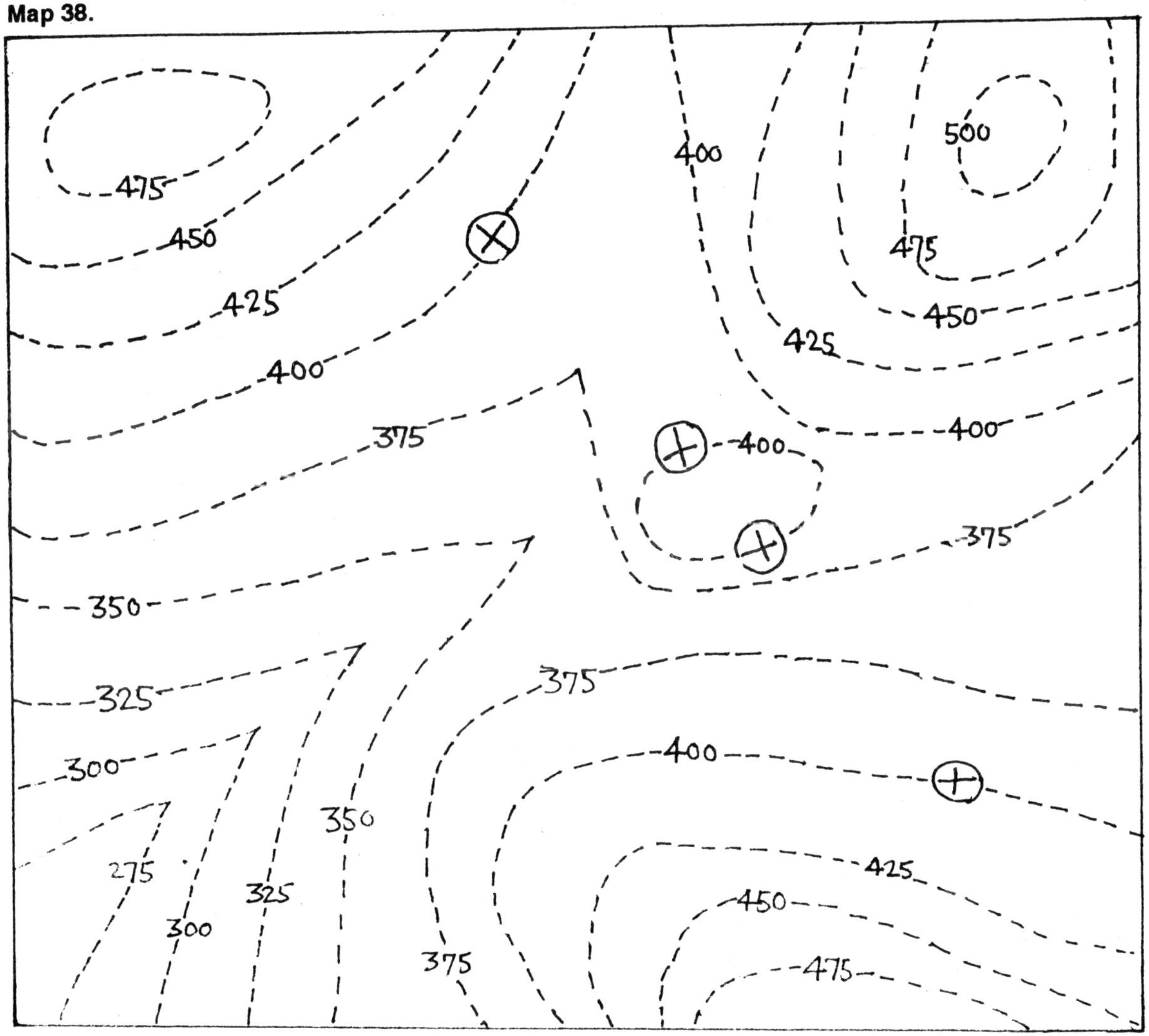

Top of a dipping sandstone bed crops out at places marked +/-. The conformable series dips 1 in 4 due northeast, in general. The thicknesses and the order of superposition of the beds, is given below.

Limistone	
Clay	100 Meters.
Coarse Sandstone	25 Meters
Sandstone	150 meters
Grit	50 Meters
Conglomerate	

Trace the outcrops of the several beds. Determine the exact direction of dip and strike of the series.

0 1 cm = 100 ut ters.

Cont. Int. = 25 Meters.

SECTION--V
DIP AND STRIKE PROBLEMS

Categories

A. Given two apparent dips, determine true dip amount, and direction of true dip and true strike.

B. Given true dip, determine dip amount in the given directions of apparent dip.

C. Given one apparent dip and true direction of strike, find true dip amount and direction.

D. Given one apparent dip, direction of true dip, find amount of true dip and strike.

E. Given one apparent dip and amount of true dip, find the direction of true dip.

F. Given true dip, to find the directions in which the given amount of apparent dip will be encountered.

G. Given one apparent dip and true dip and direction of true strike.

DIP AND STRIKE PROBLEMS

(Solve the problems by the geometric method)

Category-A

Given two apparent dips, determine true dip amount, and direction of true dip and true strike.

1. A sandstone bed has an apparent dip of 1 in 6 due north, and 1 in 10 due east. Determine the amount of true dip, and direction of true dip and true strike. (Answer-1 in 5.15 due N32°E, N58°W-S58°E).
2. On a dipping rock surface, two apparent dips have been recorded namely. i) 1 in 10 due N40°E, and ii) 1 in 10 due N10°W. Determine the amount of true dip, and direction of true dip and true strike. (Answer-1 in 9.1 due N15°30'E, N74°30'W-S74°30'E).
3. Two apparent dips were recorded on a dipping bed of sandstone, namely, i) 1 in 7.5 due S70°E, and ii) 1 in 10.5 due S20°W. Find the amount of true dip, and true strike. (Answer-1 in 7 due S34°30'E, N55°30'E-S55°30'W).
4. Apparent dips of 1 in 15.7 due S50°W, and 1 in 10.2 due N10°W were recorded on a dipping bed of shale. Find the amount of true dip, and direction of true dip and true strike. (Answer-1 in 6.2 due N62°W, N28°E-S28°W).
5. Apparent dips of 1 in 8.5 due N30°W, and 1 in 9.5 due N65°E have been recorded on a rock surface. Determine the amount of true dip, and direction of true dip and strue strike. (Answer-1 in 6.1 due N14°30'E, N75°30'W-S75°30'E).

Category-B

Given true dip, determine dip amount in the given directions of apparent dip.

6. A coal seam possesses a true dip of 1 in 4 due south. Find the amount of apparent dip due S30°W, and due S70°E. Will there be any dip amount due east or west? Give reasons for your answer. (Answer-1 in 4.6 and 1 in 11.45, respectively. No dip either due east or west because it is the direction of true strike).
7. A dipping limestone bed has a true dip of 1 in 5 due west. Find out the amount of apparent dip due N60°W, N40°W, S30°W and south. (Answer-1 in 5.8, 1 in 7.1, in 10.3 and zero respectively).
8. A dipping series of beds has a true dip of 1 in 1 due N20°E. Find the amount of apparent dip due N40°W, N70°W and N50°E. (Answer-1 in 2, zero and 1 in 1.15, respectively).
9. A dipping clay bed has a true dip of 1 in 7 due S45°W. Fine the amount of apparent dip due west, S65°W and S45°E. (Answer-1 in 10.1, 1 in 7.5 and zero, respectively).
10. A dipping quartzite bed has a true dip of 1 in 2 due south. Find the amount of apparent dip due S40°W, S50°E and east. (Answer-1 in 2.54, 1 in 3.05 and zero, respectively).

Category-C

Given one apparent dip and true direction of strike, find true dip amount and direction.

11. A shale bed has an apparent dip of 1 in 10 due N30°W. If the direction of true strike be N-S, determine the direction and amount of true dip. (Answer-1 in 5 due west).
12. A dipping bed of sandstone has an apparent dip of 1 in 10 due east. If the direction of true strike be N40°E-S40°W, find the amount and direction of true dip. (Answer-1 in 7.75 due S 50°E).
13. A dipping bed of shale possesses an apparent dip of 1 in 1 due east. If the direction of true strike be N40°W-S40E, find the direction and amount of true dip. (Answer-1 in 9.3 due N50°E).
14. A dipping bed of limestone has an apparent dip of 1 in 12 due N50°E. If the direction of true strike be N20°E-S20°W, find the direction of true dip as well as the amount. (Answer-1 in 6.1 due S70°E).
15. A dipping bed of conglomerate has an apparent dip of 1 in 8.5 due N20°W. If the direction of true strike be N-S, find the direction and amount of true dip. (Answer-1 in 2.9 due west).

Category - D

Given one apparent dip, direction of true dip, find amount of true dip and strike.

16. A limestone bed has an apparent dip of 1 in 15 due S30°W. The direction of true dip is N70°W. Determine the amount of true dip and direction of true strike. (Answer -1 in 2.2, N20°E-S20°W).
17. On a dipping bed of limestone, an apparent dip of 1 in 8.7 due S30°W has been recorded. If the direction of true dip be due S30°E, find the amount of true dip and the direction of true strike. (Answer - 1 in 4.5, N60°E-S60°W).
18. A sandstone possesses an apparent dip of 1 in 11.5 due west. If the direction of true dip be due N35°W, find the amount of true dip, and direction of true strike. (Answer -1 in 6.45, N55°E-S55°W).
19. A fault plane possesses an apparent dip of 1 in 6.5 due S75°W. If the direction of true dip be due S25°W, find the amount of true dip and the direction of true strike. (Answer-1 in 4.1, N65°W-S65°E).
20. A grit bed has an apparent dip of 1 in 6.3 due N66°E. If the direction of true dip be due N15°E, find the amount of true dip and direction of true strike. (Answer -1 in 4, N75°W-S75°E).

Gategory -E

Given one apparent dip and amount of true dip, find the directions of true dip.

21. A dipping bed has an apparent dip of 1 in 12 due S50°E. If the amount of true dip be 1 in 4, determine the directions of true dip. (Answer-S21°30'W, N58°E).

22. A dipping formation possesses an apparent dip of 1 in 10 due north. If the amount of true dip be 1 in 4.5, find the directions of true dip. (Answer-N64°E, N62°W).

23. A bed has an apparent dip of 1 in 11.5 due S20°E. If the amount of true dip be 1 in 3, find the directions of true dip. (Answer-S55°W, N83°E).

24. A dipping formation has an apparent dip of 1 in 8.5 due S80°W. If the amount of true dip be 1 in 4.7, find the directions of true dip. (Answer-S22°30'W, N43°W).

25. A dipping limestone bed has an apparent dip of 1 in 6.5 due S80°E. If the amount of true dip be 1 in 3.2, find the directions of true dip. (Answer - S17°E, N41°30', E).

Category-F

Given true dip, to find the directions in which the given amount of apparent dip will be encountered.

26. A true dip of 1 in 5 due N40°E has been recorded on a rock surface. Determine the directions in which the amount of apparent dip will be 1 in 11.3. (Answer-N24°W, S76°30'E).

27. A true dip of 1 in 2 due east has been registered on a dipping sandstone bed. In what directions the apparent dip amount will be 1 in 3.5? (Answer-S35°30'E, N35°30'E).

28. A true dip of 1 in 3.5 due S20°W, is recorded on a bed of clay. Find the directions in which the apparent dip amount will be 1 in 7. (Answer -S80°30'W, S40°E).

29. A sandstone bed has a true dip of 1 in 4 due N20°W. Determine the directions in which the apparent dip amount will be 1 in 9.5. (Answer-N84°30'W, N46°E).

30. A true dip of 1 in 3.8 due N72°E was recorded on a dipping rock surface. Determine the directions in which the apparent dip amount will be 1 in 6.7. (Answer-N8°E, S43°30'E).

Category-G

Given one apparent dip and true dip direction, determine the amount of true dip and direction of true strike.

31. A dipping sedimentary bed has an apparent dip of 1 in 10.5 due east. If the true dip be due N35°E, find the amount of true dip and direction of true strike. (Answer-1 in 6, N55°W-S55°E).

32. An apparent dip of 1 in 9.5 due north has been recorded on a rock surface. If the true dip be due N60°W, find the amount of true dip and the direction of true strike. (Answer-1 in 4.8, N30°E-S30°W).

33. **On a fault plane, an apparent dip of 1 in 5.9 due S76°W has been observed. If the direction of true dip be due S14°W, determine the amount of true dip and the direction of true strike. (Answer-1 in 2.8, N76°W-S76°E).**

34. An apparent dip of 1 in 2.6 due S47°W has been registered, on a dipping bedding plane. If the direction of true dip be due S32°E, determine the amount of true dip, and the direction of true strike. (Answer-1 in 0.55, N58°E-S58°W).

35. On a dipping series of beds, an apparent dip of 1 in 7.7 due S58°E has been recorded. If the direction of true dip be due N57°E, determine the amount of true dip, and the direction of true strike. (Answer -1 in 2.9, N33°W-S33°E).